Advances in Anatomy
Embryology and Cell Biology

Vol. 123

Editors

F. Beck, Melbourne   W. Hild, Galveston
W. Kriz, Heidelberg   J. E. Pauly, Little Rock
Y. Sano, Kyoto   T. H. Schiebler, Würzburg

K.E. Åström H.deF. Webster

The Early Development of the Neopallial Wall and Area Choroidea in Fetal Rats

A Light and Electron Microscopic Study

With 32 Figures

Springer-Verlag

Berlin Heidelberg New York
London Paris Tokyo
Hong Kong Barcelona
Budapest

Karl Erik Åström, M.D., Ph.D.
Henry de F. Webster, M.D.

Laboratory of Experimental Neuropathology NINDS,
National Institutes of Health,
Bld 36, Room 4A–29, Bethesda, MD 20892, USA

ISBN 3-540-53910-7 Springer-Verlag Berlin Heidelberg New York
ISBN 0-387-53910-7 Springer-Verlag New York Berlin Heidelberg

Typesetting: Best-Set Typesetter Ltd., Hong Kong
Printing and binding: Konrad Triltsch, Würzburg

21/3130-543210 – Printed on acid-free paper

Contents

VI

1 Introduction

As the neural tube develops it is modified locally along the neuraxis. These regional differences reflect the future organization of the CNS. In the fetal rat the anterior part of the neural tube is closed on the tenth day of gestation (E10), and develops into the forebrain (Witschi 1962). Subsequently, the telencephalon expands and forms the hemispheres. The latter, enclosing the fluid-filled ventricles, are joined in the dorsal midline by a layer of cells which is called the telencephalon medium or the telencephalic roof plate. In developing mammals, the roof-plate consists of two parts: an anterior, thicker portion, the area or lamina terminalis, and a posterior, thinner portion, the area or lamina choroidea (Bailey 1916; Warren 1917). The transition point between the two portions is called the angulus terminalis (Hines 1922). The area choroidea extends in a posterior direction from the angulus terminalis to the velum transversum, which separates the telencephalon from the diencephalon.

The *cuboidal cells* forming the neural tube are called epithelial due to their neuroepithelial origin (Cajal 1909). As the neural wall thickens, epithelial cells elongate, become bipolar and more columnar in shape; at this stage they are called *columnar cells*. The epithelium in the neural tube and wall is said to be pseudostratified since it has the appearance of being stratified but in reality is composed of a single layer of cells (Sauer 1935 a,b). Then the simple structure of the neural wall is modified when postmitotic neurons appear and migrate towards the surface of the brain. At this stage bipolar neuroepithelial cells which span the entire width of the wall are still the predominant cell type. These cells, now called *radial glial cells* (Rakic 1971, 1972), increase greatly in number and length during the period of neuronal migration.

Important studies on the light microscopic appearance of these types of neuroepithelial cells by early neuroanatomists concerned their cytology as seen in stained sections (His 1889; Sauer 1935b), and their shape and orientation as observed in Golgi preparations (Retzius 1893, 1894; Cajal 1909). The Golgi method has continued to yield useful information since it alone permits the study of neuroepithelial cells in their entirety (Åström 1967; Stensaas and Stensaas 1968; Hinds and Ruffett 1971; Schmechel and Rakic 1979).

Electron microscopy has defined important aspects of the fine structure of the neuroepithelial cells. It has shown, inter alia, that the ventricular and pial surfaces are *not* covered by membranes (which earlier anatomists claimed), that the cells are joined to each other by junctional complexes, and that they contain cytoskeletons consisting of tubules and filaments.

During development the columnar/radial glial cells change in shape, internal structure and phenotypic expression. Immunostaining methods have been employed to identify and compare their phenotypes. These methods can reveal important cell differences which supplement conventional light or electron microscope studies. Thus, immunocytochemical observations have shown that the radial glial cells contain glial fibrillary acidic protein (GFAP) in fetuses of humans (Choi and Lapham 1978) and of monkeys (Levitt and Rakic 1980). Staining for GFAP in radial glial fibers in fetuses of rodents has failed in some studies (Bignami and Dahl 1974; Pixley and DeVellis 1984) but not in others (Valentino et al. 1983; Choi 1988). However, it appears that vimentin, another intermediary filament protein, is the major cytoskeletal protein in the immature rat brain (Dahl 1981; Dahl et al. 1981). It was detected in the neural tube of fetal mice at E9 by Houle and Fedoroff (1983) and in columnar cells in the mouse brain at E11 (Schnitzer et al. 1981). More recently, Hockfield and McKay (1985) used a monoclonal antibody which stains a surface antigen on radial glial cells in fetal rats during the period of neuronal proliferation and migration, but not later. Patchy staining of columnar cells was seen as early as E11.

Up to a certain point of development, the subdivisions of the CNS retain the basic structure of the neural tube from which they originated. For example, in fetal rats, although the major subdivisions of the brain are visible on E12, the telencephalic wall at this stage of development does not contain any morphologically identifiable neurons but only what His (1889) called spongioblasts and germinal cells. This period is of fundamental importance for the subsequent development of the brain; it determines its future shape and sets the stage for the period when postmitotic developing neurons migrate to destinations in the cortical plate where they will settle, grow, and form synaptic connections with other neurons of the CNS. Although recent studies (Rickmann and Wolff 1985; Choi 1988) and reviews (Rakic 1982; Fedoroff 1986; Nowakowski 1987) have dealt with various aspects of prenatal CNS development, there are few observations on this initial period of development.

The goal of our own work in this field was to define the fine structure of cells in the telencephalic wall, their shapes, and their relationships during the preneuronal period (E11–13) in fetal rats (some specimens from E14 to 16 have also been examined for comparison). Two areas were selected for study: the lateral convexity of the hemispheric vesicle, which is called the neopallial wall, and the midline area, called the area choroidea. From the beginning of the formation of the telencephalon up to the early part of E13, both the suprastriatal portion of the telencephalon and the area choroidea retain the basic structure of neuroepithelium, i.e. each consists of a layer of columnar epithelial cells. However, differences in the shape and fine structure of the cells in these two regions are visible as soon as the telencephalic vesicles are formed.

Observations of the neopallial wall are described in Chap. 4. The purpose of this description is twofold. First, the observations provide evidence that is essential for a discussion of the nature and function of the cells which make up the subdivisions of the CNS before neurons proliferate and migrate. Our findings help show how the CNS is given its initial shape (exemplified by the telencephalon) which then is modified during development, and what mechanisms regulate the metamorphosis of dividing cells and growth of their progeny. Secondly, our observations allowed us to compare the structure of the developing neopallial wall and area choroidea. The morphology of the latter, which has not been described, previously, is presented in Chap. 5. The purpose of that chapter is to give a description of the fine anatomy of the cells in the area choroidea and to discuss their possible functions.

2 Nomenclature

Pallium is the suprastriatal portion of the cerebral vesicles.

Telencephalic wall, which is used here as a synonym for pallium, expands above the ventricular system. In the preneuronal period it is a continuous nonstratified monolayer of columnar cells.

Neopallial wall is that part of the telencephalic wall from which the isocortex will develop later.

Telencephalic roof plate is the layer of cells which joins the telencephalic hemispheres in the midline.

Area choroidea (or *lamina choroidea* or *tela choroidea*) is the posterior part of the telencephalic roof plate.

Paraphyseal arch is the posterior part of the area choroidea.

Roof cell is one cell within the telencephalic roof or area choroidea.

Columnar cell is a bipolar, neuroepithelial cell in the neopallial wall and telencephalic roof which extends from the ventricle to the pial surface. The name columnar refers only to the shape of individual cells in the early neural wall. It should not be confused with the concept of columns of neurons in the adult cerebral cortex, which are believed to be functionally related (Mountcastle 1979) or to columns of immature neurons in the cortical plate (Smart and McSherry 1982). (Nomenclature for individual parts of columnar cells is given in Fig. 27.)

Radial glial cells are columnar cells which are found in the telencephalic wall during the period of neuronal migration. Immunostaining methods reveal certain characteristic features in their cytoskeletal filaments and surface membranes.

Neuroepithelial cells. The cuboidal cells in the neural tube and both columnar and radial glial cells in the early telencephalic wall are all types of neuroepithelial cells in the sense that they emanate directly from the neuroepithelium and have the basic structure of polarized epithelial cells in general.

Apical direction is towards the ventricular lumen. *Basal* direction is towards the pial surface.

Inner process of a columnar/radial glial cell is located between the nucleus and the ventricular lumen. *Outer process* is located between the nucleus and the pial surface.

3 Material and Methods

Females rats (Sprague-Dawley) with timed pregnancies were obtained from the Charles River Farm. They were selected if a sperm plug was found the morning after mating during the preceding 12 h; 8 a.m. was considered as the time of conception although the actual gestation time might have been up to 12 h longer. The rats were kept in groups of three in large cages, and were given Purina food pellets and water ad lib. The fetuses were removed at E11, 12, or 13 (in a few cases also at E14, 15, and 16). In all, more than 100 specimens were studied.

Embryonic tissue, especially the brain, is delicate and, therefore, difficult to prepare for electron microscopy. The handling, dissection, and blocking of tissues can produce preparative artifacts; the composition of fixatives and their buffers, and methods for dehydration and embedding may also contribute to distortions. The following procedure produced light and electron microscopic preparations which were mostly free from artifacts. A detailed account of the method is given elsewhere (Åström and Webster 1990).

The pregnant mother was anesthetized by an intraperitoneal injection of chloral hydrate (30–35 mg/100 g body weight) and positioned on her right or left side. A 2-cm-long cut was made in the abdominal wall, parallel with the midline, below the edge of the thorax on the right or left side, and the fetuses were removed one at a time after ligation of blood vessels. Four fetuses were usually taken from each pregnant animal for electron microscopy. Additional fetuses were used for light microscopy and for measuring the crown-rump length.

All dissections were carried out under visual control through a dissecting microscope. The unfixed or glutaraldehyde-fixed fetuses were usually dissected in saline or fixative in the following way. The severed head, positioned on its right or left side in a fluid-filled dish, was fastened to a bottom layer of wax with insect pins thrust through the orbit and the ear opening. A cut was made with a pair of fine scissors along line 1 in Fig. 1A. The specimen was positioned crown down and held in that position by two pins, one through the rhomboid fossa into the wax and the other through the mesencephalon. Cuts were then made bilaterally along line 2. Additional cuts were made with a sharp razor blade along lines 3 and 4 in brains which had been adequately fixed in glutaraldehyde (i.e. they could be cut without distortion). In brains less firmly fixed, these cuts were delayed until 10–15 min after the specimen had been transferred from glutaraldehyde to the OsO_4 solution. The slab thus obtained was either dehydrated and embedded *in toto* or was cut into three pieces, one from each cerebral wall and one from the midline region as shown in the diagram in Fig. 1B. The pieces for microscopic examination were taken from the laterodorsal region of a

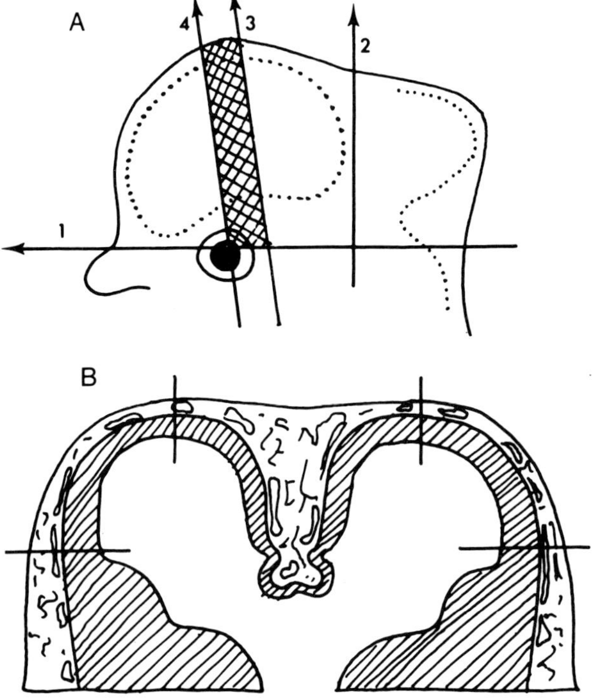

Fig. 1A,B. Methods of dissection and region selected for this study. Gestational age 13 days. **A** Whole, uncut, osmicated, transilluminated specimen, seen from the left. **B** Coronal section from midcerebral region, *hatched* in figure **A**. All illustrations are from specimens which were taken from this region. (From Åström and Webster 1990)

cerebral vesicle (the asterisk in Fig. 2B marks the approximate location) and from the area choroidea of the telencephalic roof.

The best fixation in fetuses with a gestational age of 15 days or less was achieved as follows. The telencephalon with coverings was immersed in an ice-cold 3%−4% glutaraldehyde solution containing 0.067 M s-collidine buffer and 0.3%−0.5% $CaCl_2$ for up to 30 min, and then was transferred to an ice-cold 1%−2% solution of OsO_4 in a 0.10 M s-collidine buffer containing at least 8% sucrose. A slab, cut from the mid portion of the brain after 10−15 min, was postfixed in OsO_4 for 30−50 min. Dehydration with graded alcohols started with a 50% solution; the 50% and 60% ethanol solutions contained 0.25% NaCl to prevent swelling.

Gelatine capsules or rubber molds were used to embed the tissues in epon. Semithin sections, 2−3 μm in thickness, were cut with glass knives on a Porter-Blum MT-1 microtome or an LKB Ultrotome III and stained with paraphenylenediamine. Thin sections, cut from selected areas with diamond knives, were stained for approximately 30 min in a 4%−8% solution of uranyl acetate dissolved in 1% methanol and for 15−60 s in a 0.2% aqueous solution of lead citrate. By using the "mesa technique", sections could be taken from different regions of the same coronal slab. The sections were examined in a Siemens Elmiskop I or a Philips 300 electron microscope.

4 Neopallial Wall

Starting on E11 the telencephalon grew rapidly, forming first a single cavity and then two evaginated lateral ventricles (Fig. 2). At E12 the major subdivisions of the cerebrum including the hemispheres, lateral ventricles, foramina of Monroi, interhemispherical fissure, and anlage of the choroid plexus were visible. However, at this stage of development the neopallial wall and the telencephalic roof were composed of a continuous sheet of columnar cells (Fig. 2B, C) which still had the simple structure of the epithelium forming the neural tube. Morphologically recognizable neurons had not yet appeared.

At early stages of development (E11, 12) the nuclei of the columnar cells were evenly distributed across the pseudostratified epithelium of the neural wall (Fig. 3A). These nuclei and their perikaryal cytoplasm constituted the main part of the wall, which was called the ventricular zone by the Boulder Committee (1970).

The outermost part of the wall, the marginal zone, contained the distal processes of the columnar cells but was free of nuclei. The early neural wall had only two zones; there was no mantle zone or cortical plate as yet. In this period (E11–13) the neopallial wall grew in thickness and surface area due to the elongation of the columnar cells and the addition of new cells. The approximate length of the columnar cells within the region we studied was as follows: E12, 95 µm; E13, 120 µm; E14, 140 µm.

Later during E12 or early E13, rounded and oval cells appeared in the marginal zone (Figs. 3B, 4). These were the first cells in the telencephalic wall that could be identified morphologically as neurons. After E13 the columnar cells were replaced by radial glial cells. The pseudostratified appearance of the neopallium was gradually obscured due to the migration of neurons and the subsequent formation of the cortical plate; it was no longer apparent after the migration of nerve cells had been completed.

A typical columnar cell could be seen *in toto* only in a Golgi preparation (Fig. 3A). It was widest in the perinuclear region from which the cytoplasm tapered off as the thin inner and outer processes. These cells were radially oriented throughout the telencephalic wall although processes of an individual cell rarely took a straight course. Due to irregular thickenings (varicosities), which were seen in Golgi preparations and electron micrographs (Fig. 8C), the diameter of the radial fibers was not uniform. The inner processes, which usually were somewhat thicker than the outer ones (diameter 1–1.5 µm) reached the ventricular surface, and the outer ones (diameter 0.7–1.0 µm) extended to the pial surface.

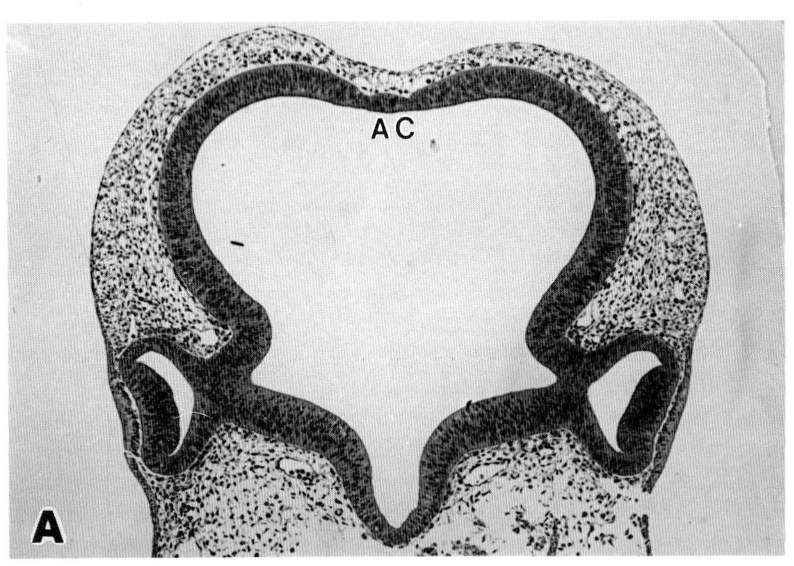
AC

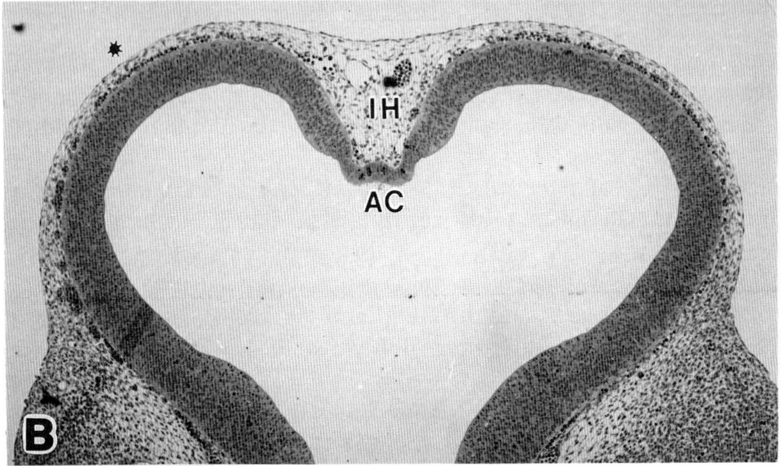
IH
AC

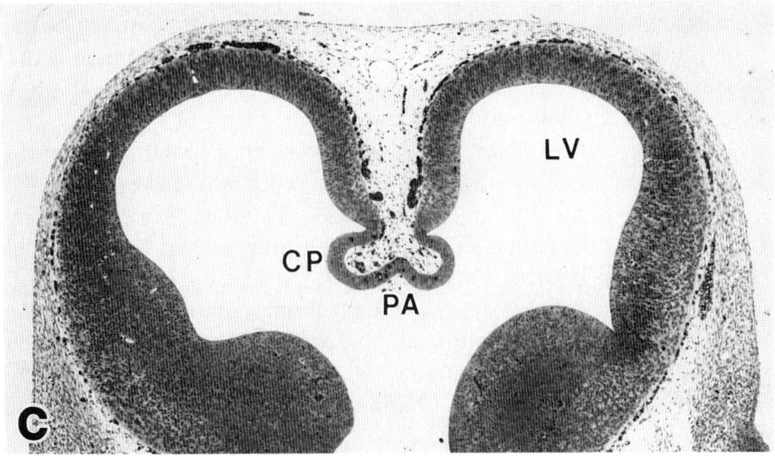
LV
CP
PA

In the preneuronal period the typical bipolar shape of the columnar cells was only seen in those cells that had their nuclei in the middle of the telencephalic wall (Figs. 3A, 27C). Cells with nuclei near the pial and ventricular surfaces had plump outer and inner processes respectively (Figs. 5, 11A).

Cytological examination, especially in the electron microscope, showed that the columnar cells were asymmetrical and had regions with different morphological features. Starting from the ventricular surface these regions were: apical (ventricular) end, inner process, nucleus and perikaryon, outer process, and terminal (basal) region with end-feet on the basal lamina (Fig. 27C).

4.1 Perikarya and Nuclei

The nuclei were oval and had finely dispersed chromatin. Numerous clusters of ribosomes were evenly distributed in the perinuclear cytoplasm (Fig. 5). They were more sparse in "attenuated" cells (Fig. 6). Perikaryal regions also contained microtubules, mitochondria, sparse microfilaments, and short cisternae of rough endoplasmic reticulum (RER) in small numbers, but no smooth endoplasmic reticulum (SER) membranes and usually no Golgi profiles.

◀ **Fig. 2A–C.** Phase micrographs of coronal sections through forebrain of fetal rats at E11, E12 and E13. The neopallial wall and the roof plate are made up of a monolayer of cells. The brains are surrounded by connective tissue but leptomeninges have not yet been formed. Blood vessels are seen at the surface of the hemispheres and in the interhemispheric area in C. The neopallial walls (but not the lamina choroidea) also contain small vessels which appear as unstained spots in C. **A** The section from a fetus at E11 is at the level of the developing eyes. The anlage of the two hemispheres are joined in the dorsal midline by area choroidea (*AC*). Forebrain has only one single ventricular cavity at this stage of development. ×50. **B** Area choroidea (*AC*) is clearly demarcated from the expanding hemispheric walls in a fetus at E12. An interhemispheric region (*IH*) has appeared. Sections for study of the neopallial wall were taken from the laterodorsal region of the telencephalic wall (*). ×40. **C** – Two hemispheres with lateral ventricles (*LV*) have been formed, and the interhemispheric region has deepened in a fetus at E13. The choroid plexus (*CP*) has started to develop at the lateral edges of the choroid area. The paraphyseal arch (*PA*) is formed through invagination of the choroid lamina: a smaller indentation is also seen in **B**. ×30

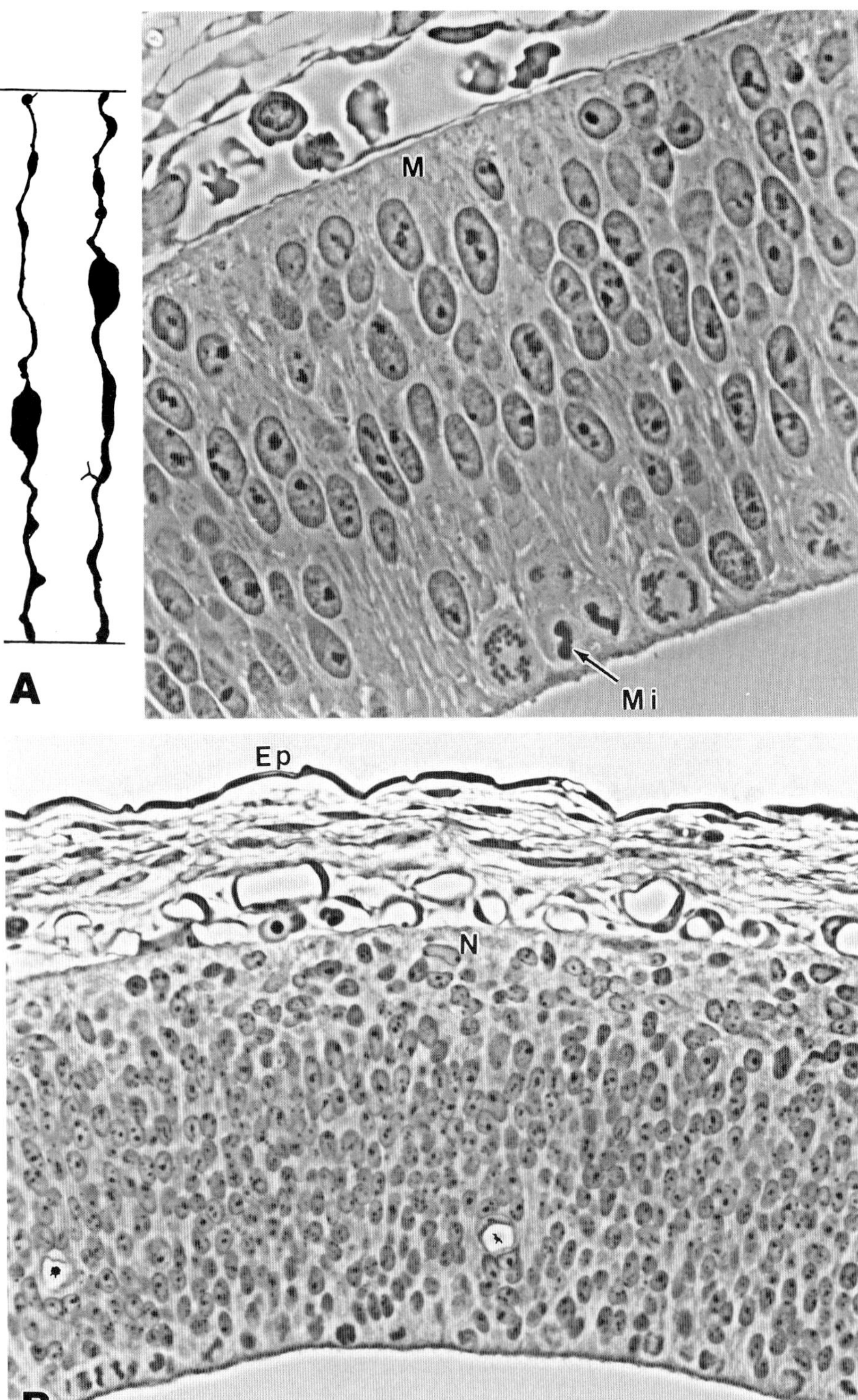
A
M
Mi
Ep
N
B

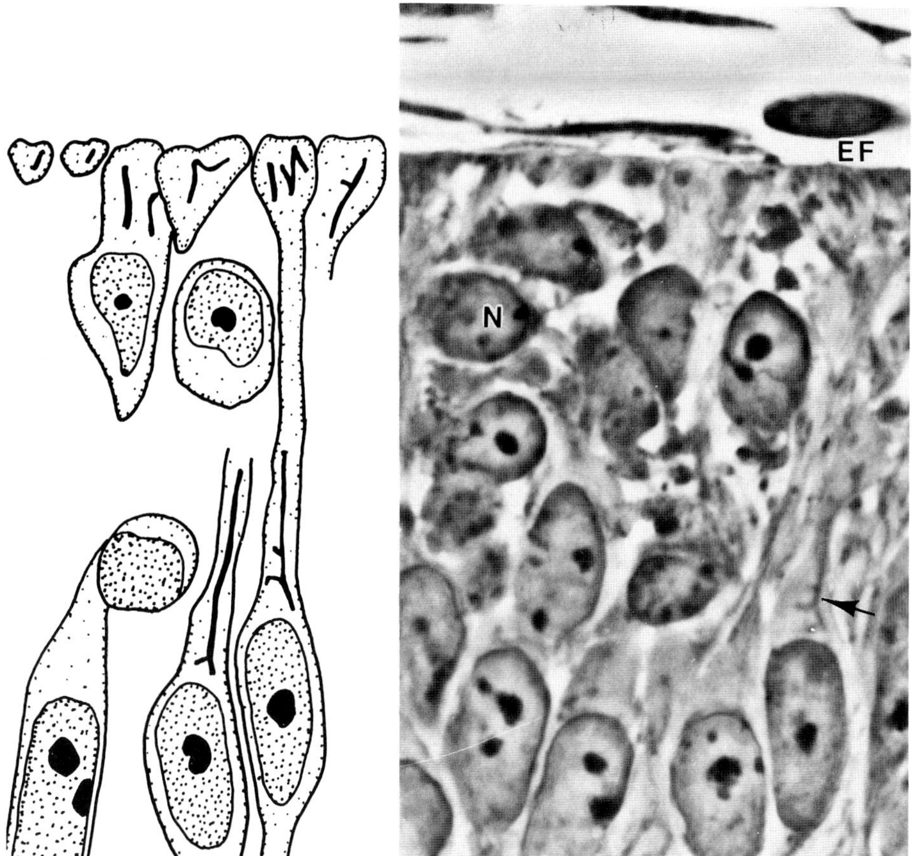

Fig. 4. Phase micrograph of the outer region of the cerebral wall in a fetus at E13. The drawing on the *left* clarifies the appearance of some cells. Columnar cells in the *lower part* of the picture have processes which terminate in club-like end-feet (*EF*) at the pial surface. Immature, rounded neurons (*N*) without processes are interposed between the layer of end-knobs and the ventricular zone which is made up of nuclei and perikarya of the columnar cells. The outer processes contain thread-like mitochondria (*arrow*). ×2000

◀ **Fig. 3. A** Phase micrograph showing the pseudostratified epithelium of the telencephalic wall in a fetus at E12. The ventricular cavity is in the *right lower corner*. Loose connective tissue with blood vessels covers the pial surface in the *left upper part* of the picture. *M*, marginal zone; *Mi*, mitotic cell. The shape of two columnar cells at E12 is shown in a drawing from a Golgi preparation. ×840. **B** Phase micrograph showing the neopallial wall in a fetus at E13. Brain is surrounded by mesenchymal tissue which is covered by epithelium (*Ep*). Nerve cells (*N*) have appeared in the marginal zone. Note blood vessels (***) within the wall (not seen in **A**). ×400 In this and all following illustrations, except for Fig. 21, the pictures are oriented so that the apical parts of cells are oriented downwards and basal parts upwards

11

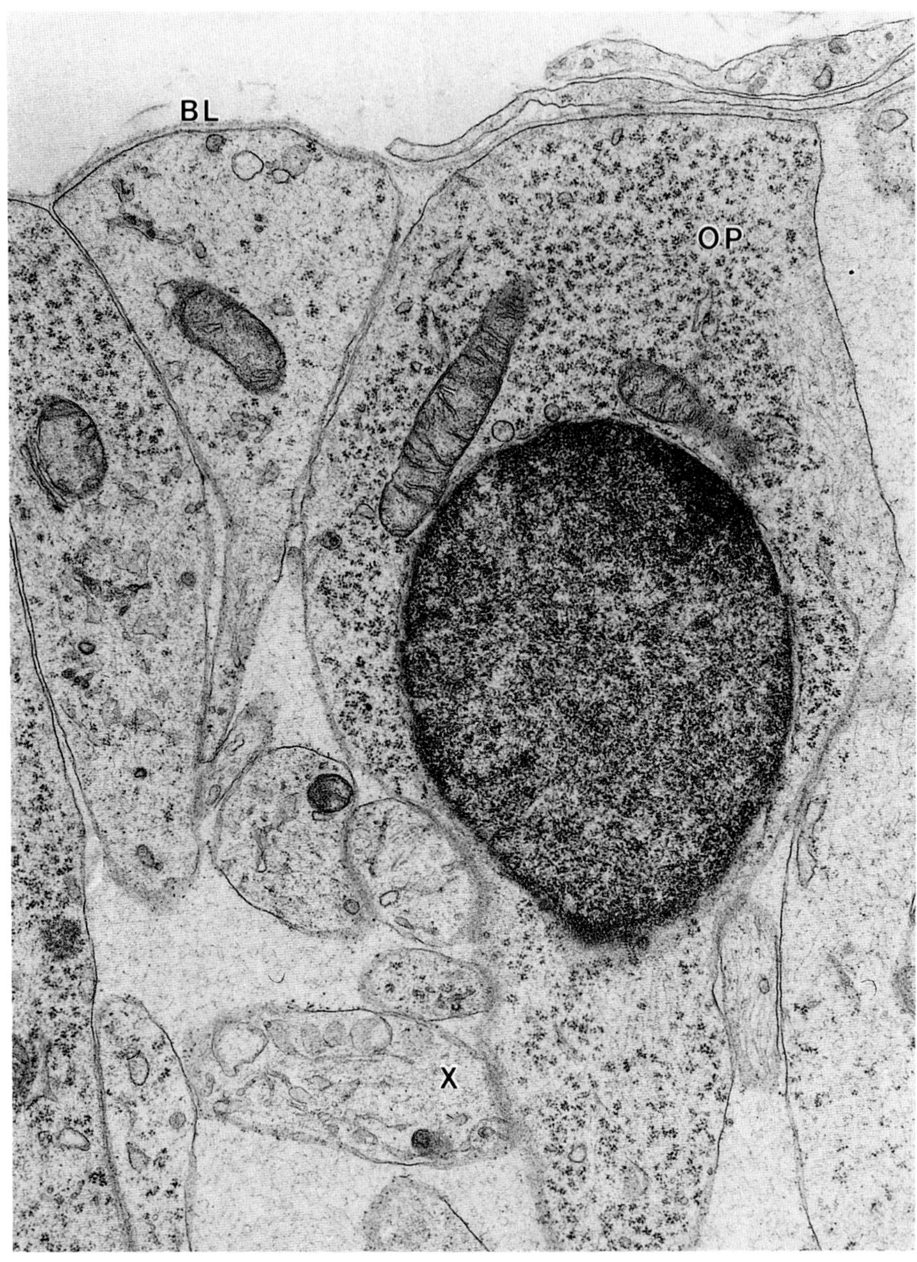

Fig. 5. E14 columnar cell with nucleus near the pial surface is wedge shaped. The outer process (*OP*) is short, plump, and flattened against the inside of the basal lamina (*BL*). Perinuclear cytoplasm contains rosettes of ribosomes but no profiles of RER. Intercellular spaces are wide. X, horizontally oriented process thought to belong to a neuron. ×18000

Fig. 6A,B. At E13 an outer process is attenuated when the nucleus is further away from the ► pial surface; compare **B** with **A** and both with Fig. 5. The concentration of polyribosomes decreases in attenuated processes. Note microtubules (*MT*) in **B**. *BL*, basal lamina. ×16300 **A**; ×16500 **B**

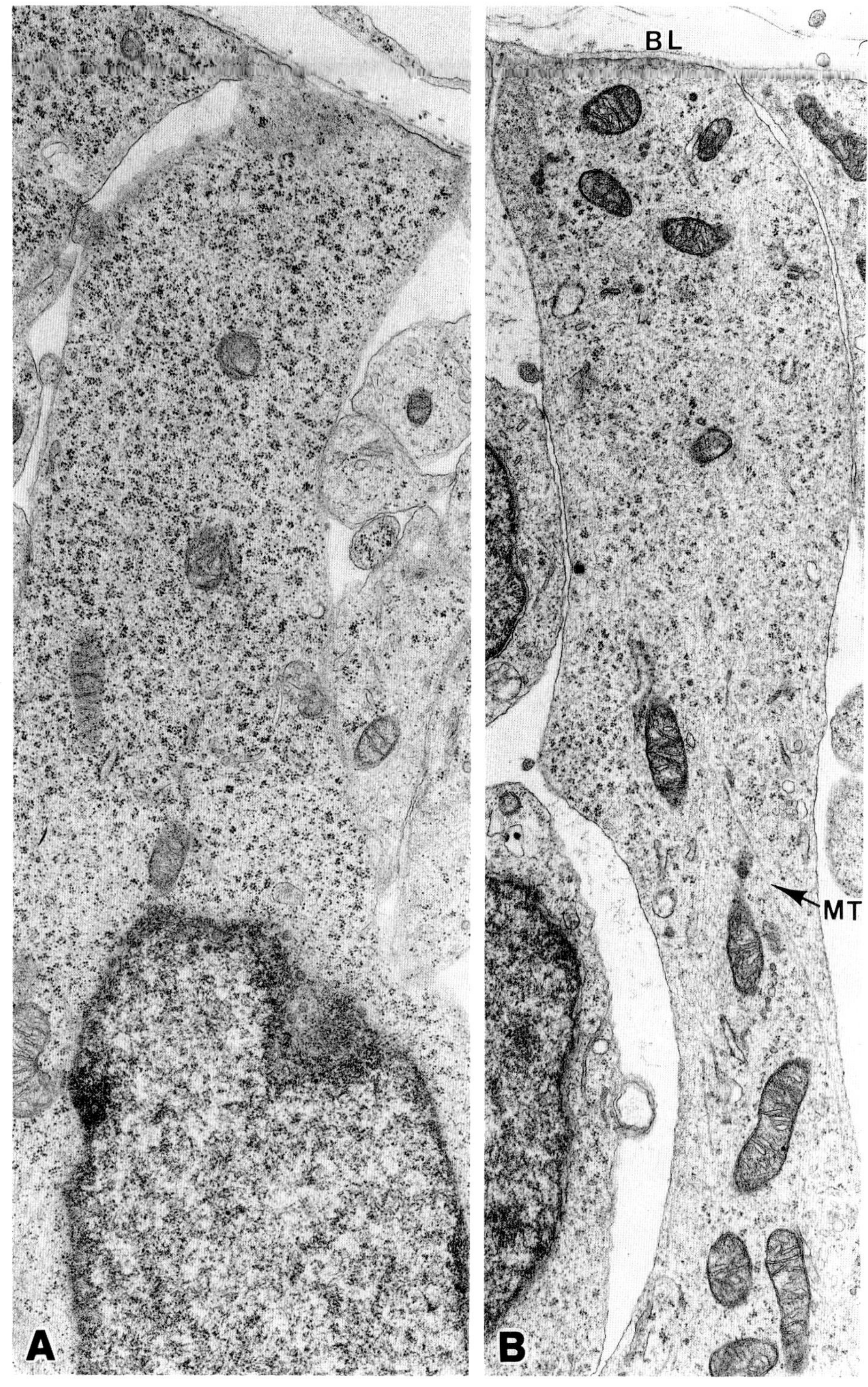

13

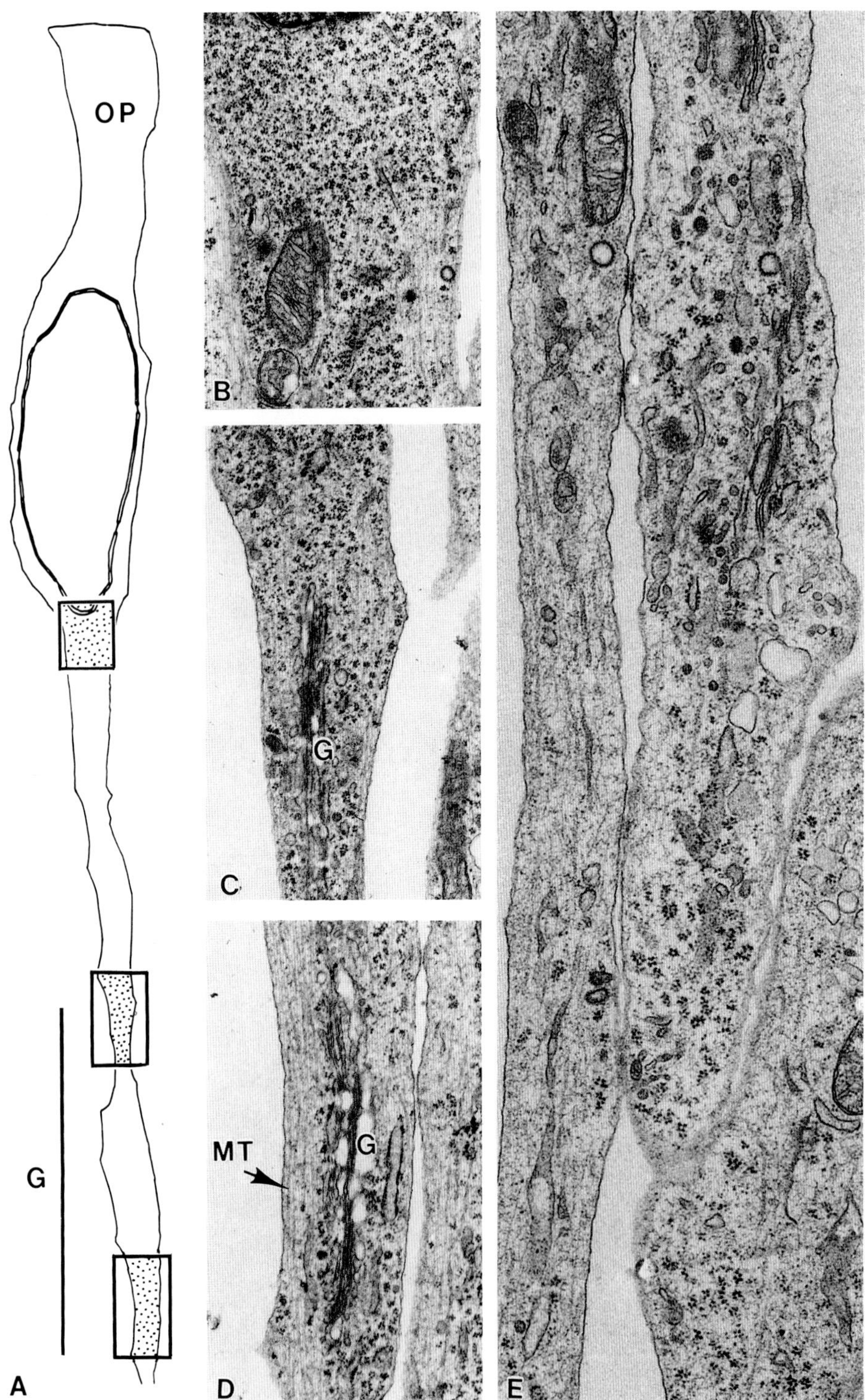

14

4.2 Inner and Outer Processes

The inner and outer processes (radial fibers) were similar but not identical in electron microscopic appearance. Microtubules with a diameter of 15–20 nm were seen in all parts of the columnar cells except the distal part of the outer processes (Figs. 7D, 8B, 9). They were situated close together in the inner and outer processes but were more dispersed in the perikarya. Most of them had a longitudinal orientation. Axially oriented profiles of SER were observed in the radial processes (Figs. 7E, 8A) but *not* in the perikarya. The processes also contained threadlike mitochondria with short branches (Fig. 4). Diffuse webs of microfilaments (Fig. 8A) were best seen in regions which lacked ribosomes. The varicose-like thickenings of cytoplasm seen in Golgi preparations (Fig. 3A) contained ribosomes (Fig. 8C) and occasionally short cisternae of RER but no microtubules or SER were seen.

In cells with nuclei near the ventricular surface (Fig. 11A) the Golgi apparatus was situated in the perikaryon near the inner pole of the nucleus or extended from there into the inner process or, less commonly, was situated beside the nucleus. In a bipolar columnar cell it was located in the inner process at some distance from the perikaryon. It was narrow, elongated, and was made up of stacks of axially oriented saccules. For instance, in Fig. 7A it has a total length of about 14 μm and its outer end is located about 15 μm from the inner pole of the nucleus. It should be emphasized that a Golgi apparatus was *never* seen in the outer processes of columnar cells.

◄ **Fig. 7A–E.** Contents of E12 radiating processes. **A** Reconstruction of a columnar cell from overlapping electron micrographs, three of which are shown. The outer process (*OP*) is wide since the nucleus is situated near the pial surface. The lower end of the inner process in the drawing is equidistant between the pial and ventricular surfaces. The *bar G* on the left side shows the location and length of the Golgi apparatus. **B–D** Appearance of the inner process within the *rectangular regions*. Note especially the elongated Golgi apparatus (*G*) and the microtubules (*MT*). ×16500. **E** Two inner radiating processes. The left process contains microtubules and SER. The right process contains parts of a Golgi apparatus, secretory vesicles, SER, polyribosomes, one coated vesicle, and some other vesicles. ×23500

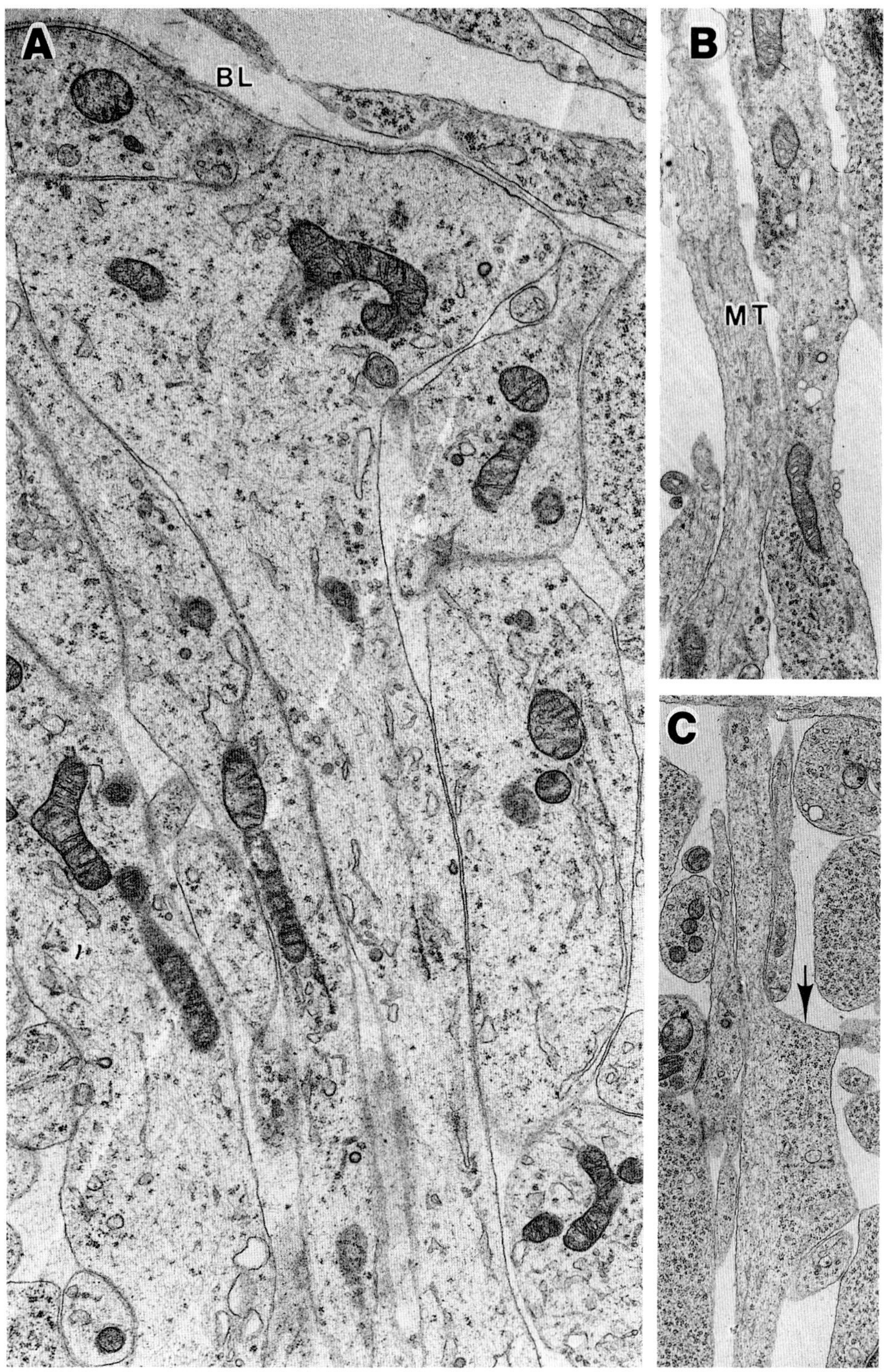
A
BL
B
MT
C

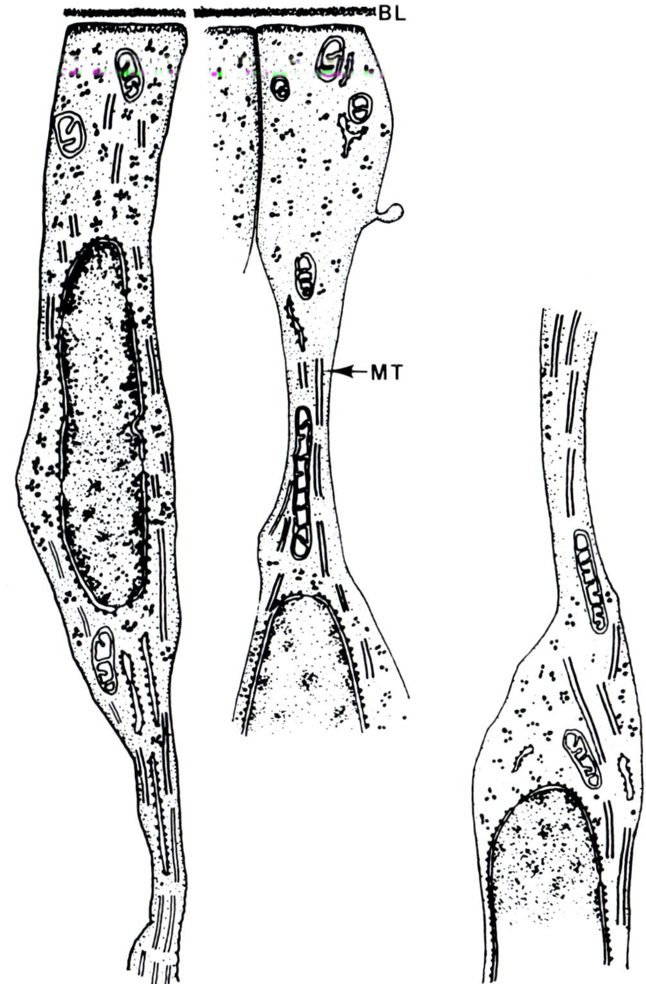

Fig. 9. In order to show the location of microtubules (*MT*) at E12 clearly, tracings were made of electron micrographs taken at low magnification. The nuclei are situated at different distances from the pial surface. *BL*, basal lamina

◀ **Fig. 8. A** The outer process of an E13 columnar cell terminates in an end-foot at the pial surface. It contains microfilaments, which form a thin web beneath the basal lamina (*BL*), SER, mitochondria, and vesicles, but few ribosomes. ×13800. **B** The thin outer radiating processes of E12 bipolar cells contain microtubules (*MT*). ×10640. **C** Varicosities on radiating process (*arrow*) of an E13 cell contain polyribosomes. ×10600

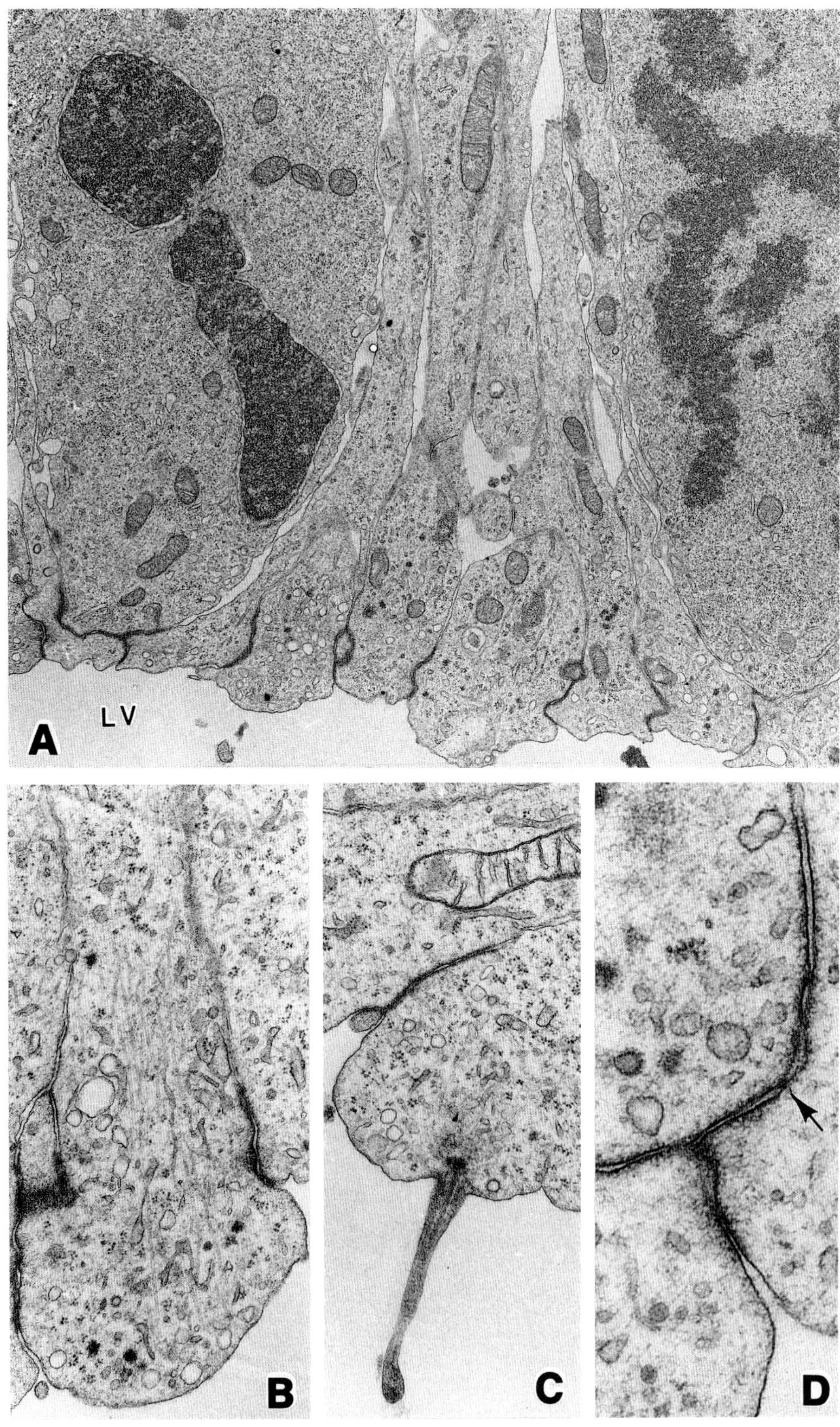

LV
A
B
C
D

4.3 Ventricular (Apical) Ends

The inner processes terminated in thickened ends, which faced the ventricular cavity and were attached both to each other and to dividing cells by junctional complexes that formed a mosaic (Fig. 10). The ventricular surfaces of these processes were usually flat or only slightly elevated above the junctional zones (terminal bars) but some protruded into the ventricular cavity (Fig. 10B) as hemispheric, folded, or finger-like processes. Cilia were rare at this stage (Fig. 10C); they were more numerous later in development. A small number of fine, short filaments projected from some parts of the ventricular surfaces (Fig. 10D). Microfilaments formed a collar around the inner ends of the cells at the level of the junctional complexes.

Some axially oriented microtubules extended from the inner radial processes into their apical ends (Fig. 10B). Less commonly, some ends contained tubules which were parallel with the ventricular surface. They were probably remnants of mitotic spindles. The ventricular ends of some bipolar cells also contained profiles of SER and small numbers of isolated vesicles. Micropinocytotic vesicles were rarely seen here.

4.4 Pial (Basal) Ends

Cells with nuclei near the *pial* surface were wedge shaped (Figs. 5, 27A). The outer process of these cells was plump and flattened against the inside of the basal lamina; only the inner process had the typical slender appearance of a radial fiber. In cells with nuclei at a somewhat deeper level, the outer process was thinner and, of course, longer since it was still in contact with the pial surface (Figs. 6A, 7, 27B). As the nucleus "moved" away from the pial surface the outer process became increasingly attenuated (Fig. 6B) until it had the typical appearance of a radial fiber. Wedge-shaped cells with nuclei near the pial surface were common at an early stage of development (E11), decreased in number during the following two days, and were not seen after E14.

◀ **Fig. 10A–D.** Ventricular aspect of the E13 neopallial wall. A Slender inner processes of columnar cells are interposed between two mitotic cells. All cells are joined by junctional complexes which, together with webs of microfilaments, appear as terminal bars in the light microscope. *LV,* lateral ventricle. ×12900. **B** The end of an inner process, containing microtubules, SER, and a few vesicles, protrudes into the ventricular cavity. ×20400. **C** A cilium is shown; at this stage of development, cilia are rarely seen. ×20400. **D** Intercellular spaces at the level of the junctional zone contains electron-dense material. The space is completely obliterated in spots (*arrow*). ×64400

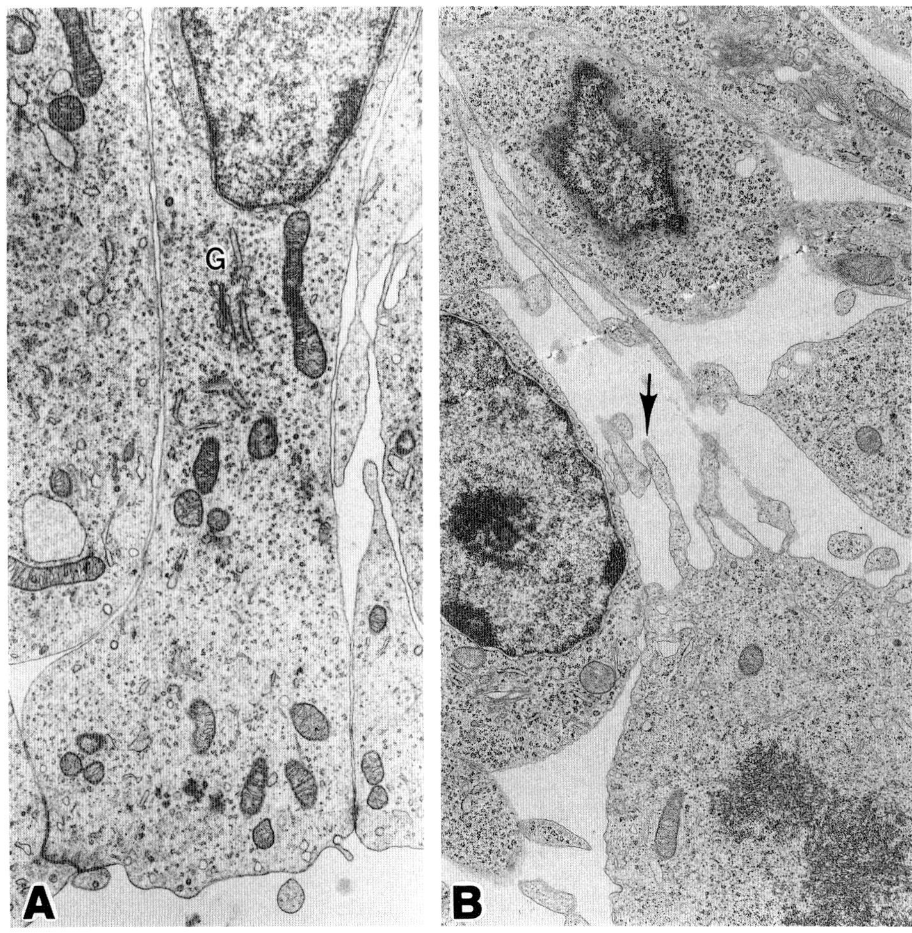

Fig. 11A,B. Ventricular region. **A** An E12 cell with nucleus near the ventricular surface has a plump inner process which contains mitochondria and polyribosomes. A small Golgi apparatus (G) is seen below the nucleus. ×8700 **B** Tail-like processes (*arrow*) project from the outer pole of mitotic cell. Intercellular spaces are wide. E11, ×9100

Fig. 12A–E. Vesicles in end-feet at E12. **A** – Groups of small vesicles may coalesce to form ▶ larger ones. ×17000. **B** Small vesicles appear to be formed from SER. ×20400. **C** A large vesicle (*arrow*) is in communication with the extracellular space beneath the basal lamina. ×34500. **D** The surface membrane probably enlarges when vesicles merge with it. ×34500. **E** Small and large vesicles in two end-feet. ×34000

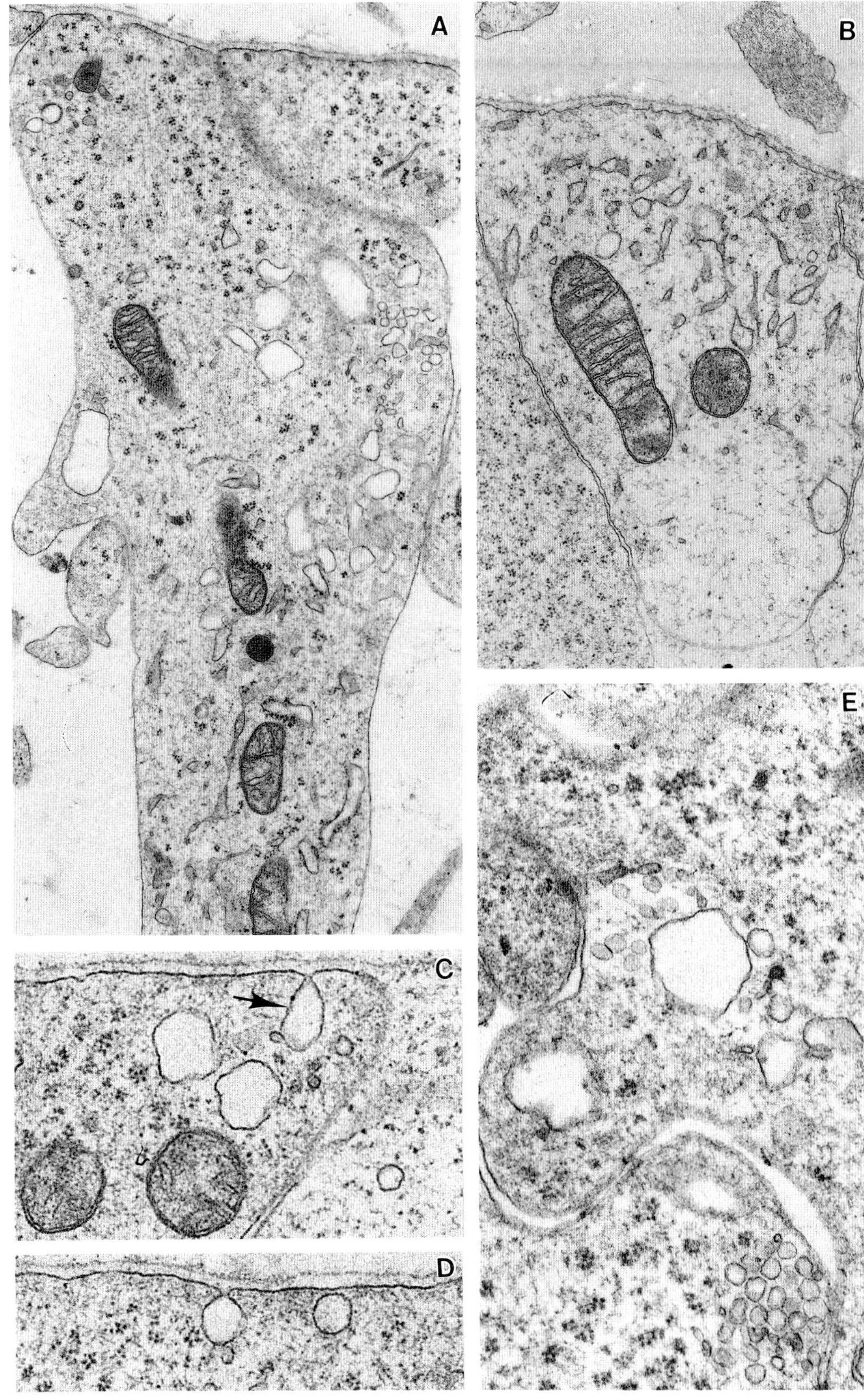

In the early preneural period each outer process terminated in a thickened ending, a so-called end-foot, in which the outer surface (facing the basal lamina) was flat (Figs. 8A, 12A). These end-feet contained ribosomes which were sparse (compared to those in the perinuclear cytoplasm), mitochondria, and SER (Fig. 12A, B). Microfilaments formed a thin mat beneath the pial surface membrane and a more diffuse web in the cytoplasm (Fig. 8A). Vesicles were prominent in some end-feet. Fig. 12 illustrates two types: (a) Smaller vesicles, which had a diameter of 0.03–0.07 µm, were round, contained electron dense material, occurred in clusters, and were frequently located at the side of an end-foot (Fig. 12A, E). These small vesicles probably originated from the tubules of SER in the end-feet (Fig. 12B). (b) Larger vesicles, which had a diameter of 0.08–0.4 µm, were round, oval or more irregular in shape; they occurred singly or in groups of two to four, did not contain any electron dense material, and had no predominant location in the end-feet (Fig. 12A, B, C). The small vesicles occurred alone or together with the large ones. Confluence of small vesicles probably gave rise to the larger ones (Fig. 12E).

Some large vesicles had openings at the cell surface (Fig. 12C, D). They were not coated and, therefore, different from those vesicles that take part in the process of micropinocytosis in the area choroidea (see below). The end-feet also contained rare multivesicular bodies and coated vesicles.

Fig. 13A–C. Pial surface at E13. **A** The pial surface is covered by loose connective tissue with ▶ one blood vessel (*BV*). Profiles of end-knobs have a flat base at the surface and a conical shape. The end-knobs form an external limiting layer of cytoplasm which has occasional gaps (*arrowhead*). Intercellular spaces below this layer are comparatively wide. *BL*, basal lamina. ×11000. **B** Cytosegregation in end-knob; the segregated area contains more polyribosomes than the cytoplasm that surrounds it. ×17400. **C** Endknob contains profile of SER (*arrow*). ×17400

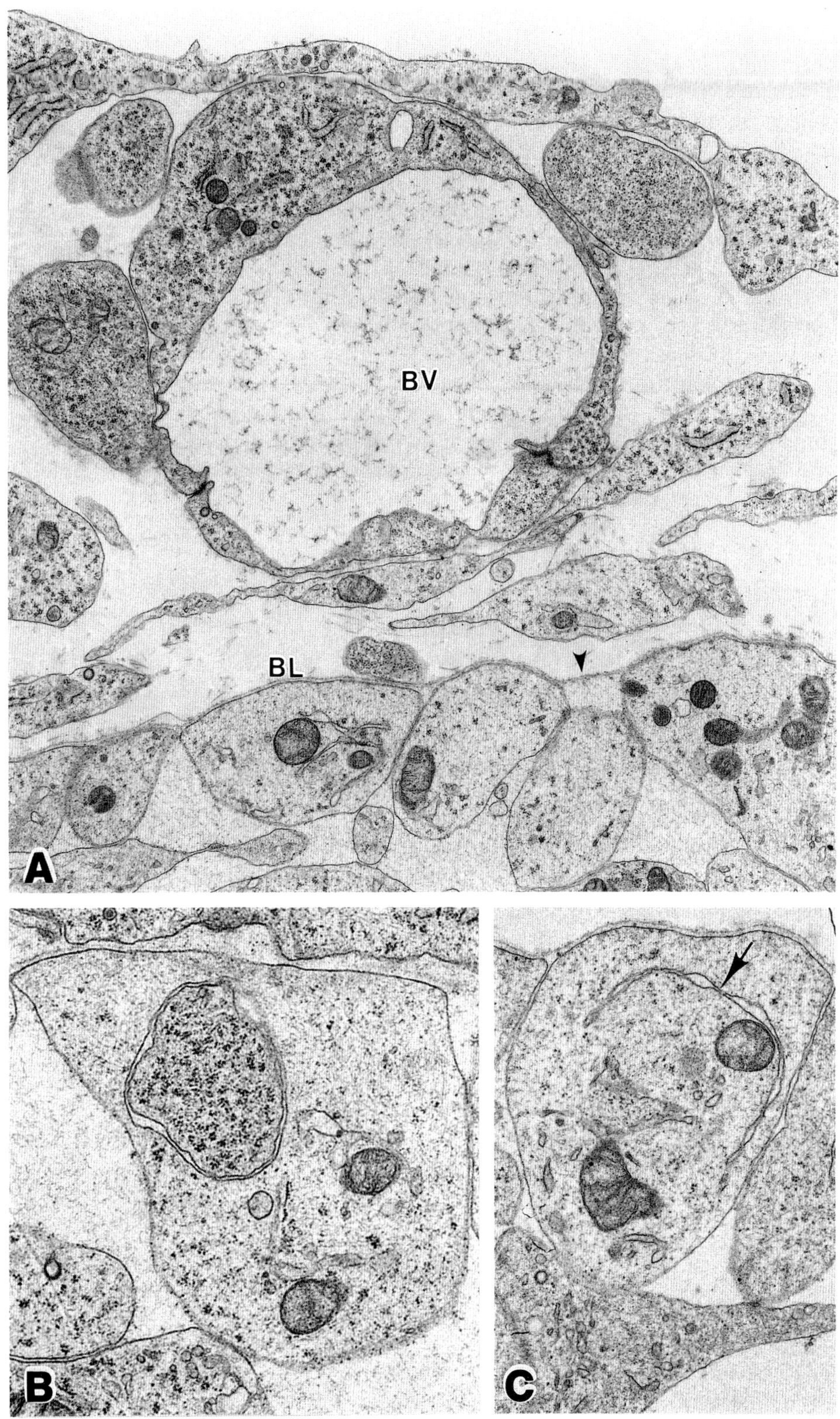
BV
BL
A
B
C

During the preneural period the outer processes began to divide distally and form branches which terminated in enlargements, called end-knobs. In the light microscope, the end-knobs appeared to form a continuous layer of cytoplasm beneath the basal lamina, but electron microscopic examination showed occasional gaps in this layer (Fig. 13A). They had a flat base at the pial surface and a conical shape (Figs. 13, 14, 15A) and were usually closely apposed but not joined together by junctional complexes. A scarcity of ribosomes and organelles gave the end-knobs a "watery" appearance.

Some of the end-knobs contained profiles of SER (Figs. 12B, 13C). It is likely that these profiles are interconnected *in vivo* since some of the profiles had bifurcations. This anastomosing network probably communicates with the axial system of SER in the outer processes.

Areas of cytosegregation with autophagia were seen in some end-knobs (Fig. 13B). Of particular interest in this figure is the difference in organelle content between the enclosed area and that surrounding it. Cytosegregation was rare in other parts of the bipolar cells in the neopallial wall, but was common in the telencephalic roof cells.

Fig. 14. At E13 rounded nerve cells (*N*) are located below the external limiting layer of ▶ end-knobs. Intercellular spaces are wide. Parts of two red blood corpuscles are seen within a blood vessel (*BV*) in the *upper left corner*. *BL*, basal lamina. ×9900

24

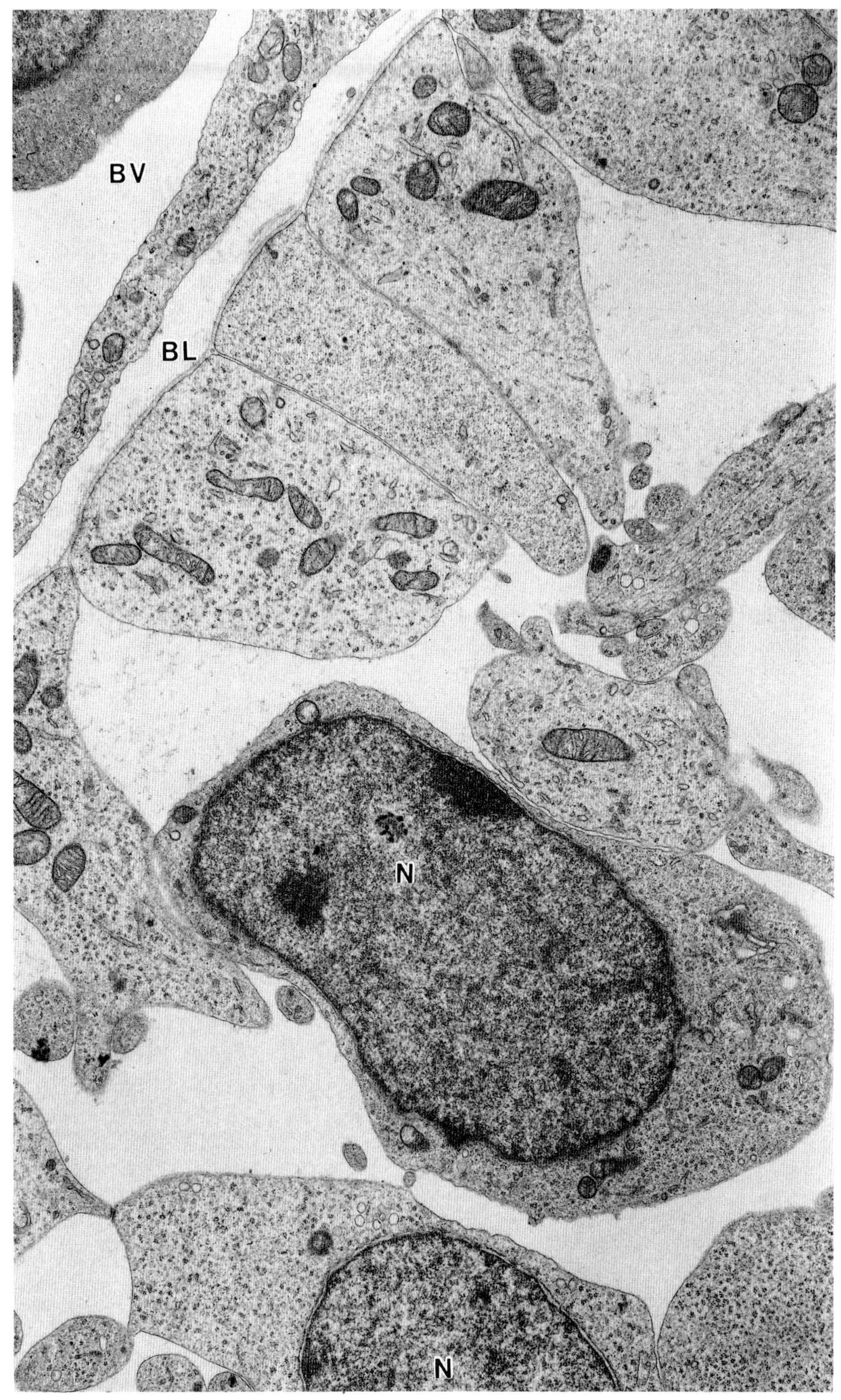

BV
BL
N
N

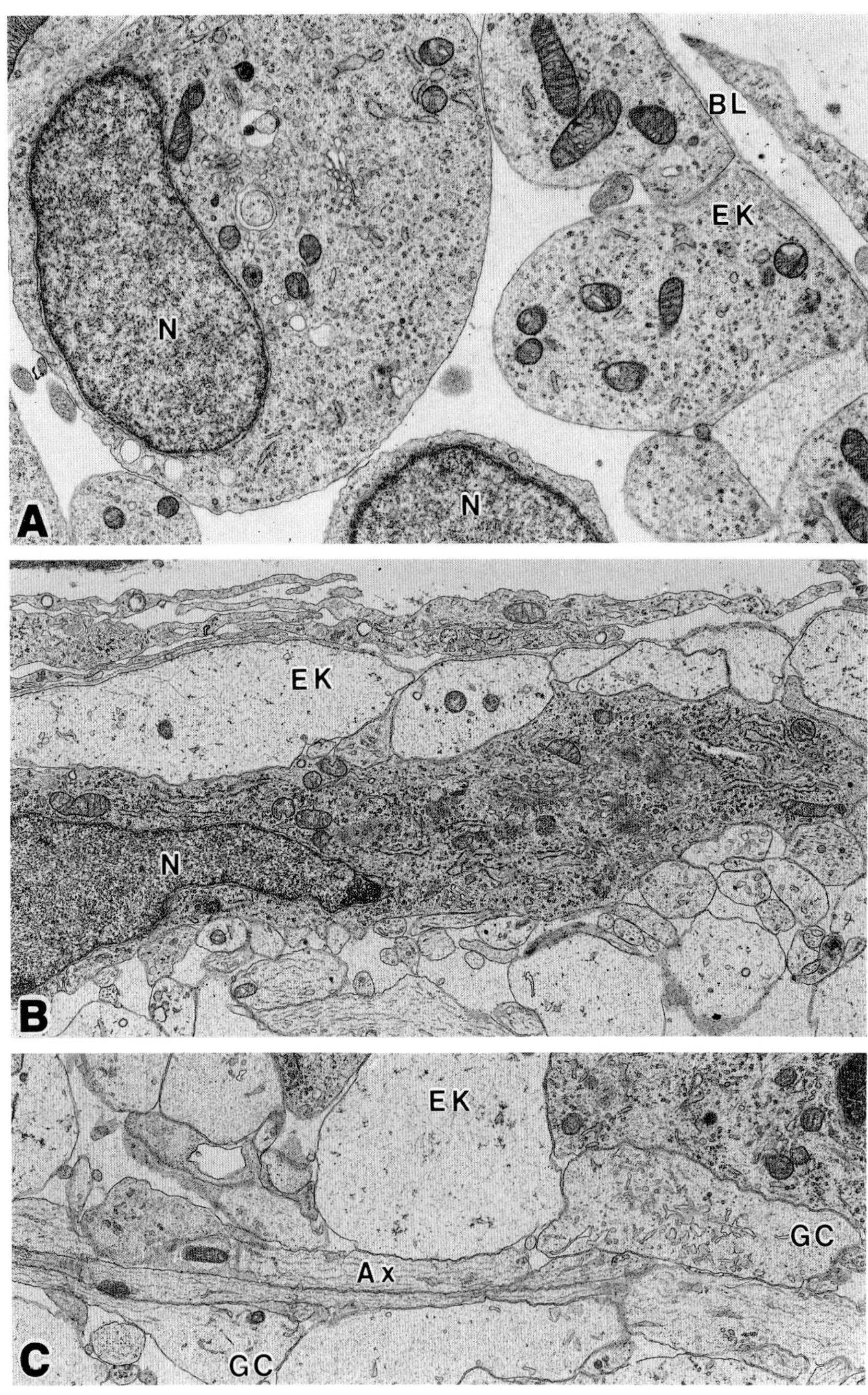
A
BL
EK
N
N
B
EK
N
C
EK
GC
Ax
GC

A low-density space measuring 50–100 nm separated the end-feet from the basal lamina, which had an average thickness of 6 nm, was dense and amorphous, and was covered on the outside by a thin layer of collagenous and reticular fibrils.

The development of the neopallial wall during the period of neuronal migration is not described in this paper. Still, some features of the radial glial cells and of the earliest recognizable neurons will be described here since they are pertinent to our description and discussion of columnar cell development.

Golgi preparations in an earlier study (Åström 1967, Figs. 4A, B, and Fig. 8) have shown that most but not all radial glial cells traverse the telencephalic wall, that bouquets of distal branches with end-knobs are present not only at the pial surface but also within the wall, and that the distal branching of the outer portion of a radial glial cell increases during development. These findings are also illustrated in Fig. 32 in this paper. The electron microscopic findings in this report showed that the end-knobs of radial glial cells that did not reach the surface usually were spherical (Fig. 16 EK1; cf Fig. 32e). On the other hand, the end-knobs of cells that traversed the entire wall had a flat base facing the surface and a triangular or more irregular shape in profile (Figs. 16 EK2 and 17 EK; cf Fig. 32f).

◀ **Fig. 15A–C.** Micrographs show a growth cone and the appearance of end-knobs at different stages of development. ×7700. **A** Immature rounded nerve cells in marginal zone are located beneath the superficial layer of end-knobs at E14. *N*, neuron. *EK*, end knob. *BL*, basal lamina. ×11400 **B** Neurons have developed horizontally oriented processes which are covered by cytoplasmic layer of end-knobs at E16. Comparisons between **A** and **B** show that the end-knobs at this stage of development are more translucent and less rich in organelles than those at E14. ×9500. **C** Three horizontally oriented immature axons terminate in growth cones (*GC*) at E16. Compare differences in structure of growth cones and end-knobs (*EK*). *Ax*, axon. ×9400

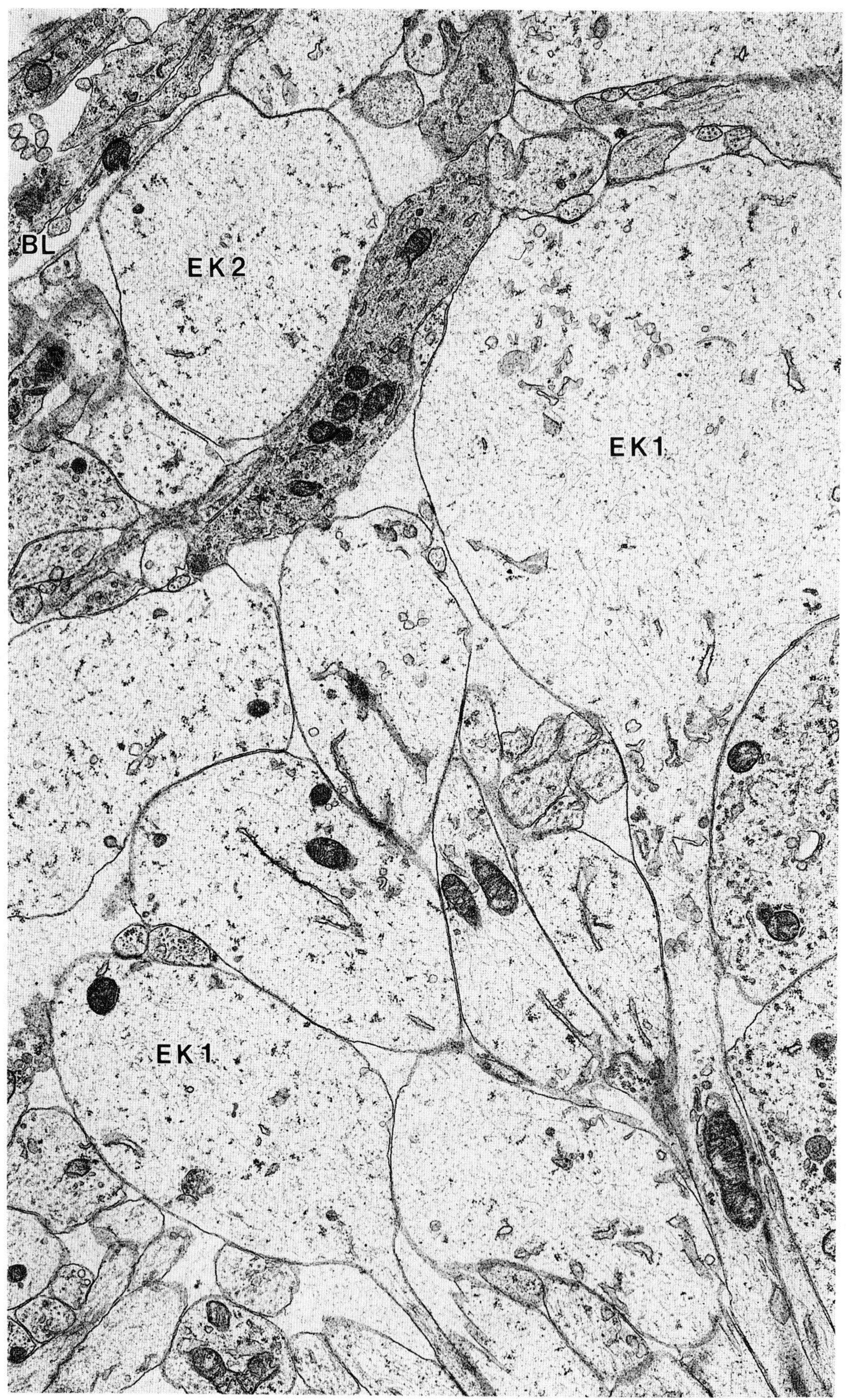
BL
EK2
EK1
EK1

Radial glial cells and columnar cells had the same basic morphology but differed in some respects. For instance, in radial glial cells the peripheral branching was more extensive and the end-knobs were larger, more electron lucent, and had fewer ribosomes (Figs. 15–17). The end-knobs in columnar cells and radial glial cells resembled growth cones of neurites in that they were the enlarged endings of developing processes and contained few organelles except for profiles of SER. The main morphological difference was that the end-knobs of columnar/radial glial cells were larger and more electron-lucent than the growth cones of neurites. They also had a different orientation than neuronal growth cones (Fig. 15C).

4.5 Relations Between Columnar Cells

At the level of the apical terminal bars, the space between neighboring cells usually was 5–18 nm; it contained some electron-dense, fuzzy material, but was completely obliterated in spots (Fig. 10D). Elsewhere along the length of columnar cells, intercellular distances were more variable. The cells were closer together in the middle part of the wall. Near the ventricle (see *e.g.* Figs. 3A, 10A) and near the pial surface (Figs. 4, 14) extracellular spaces were larger. Some faint, electron dense material was occasionally seen in areas, where the surface membranes of two neighboring cells were almost in contact. The intercellular spaces elsewhere outside the junctional zones contained no electron dense material. Distances between end-knobs were usually around 15–20 nm; junctional complexes were not seen here.

4.6 Mitotic Cells

The dividing cells were located along the ventricular cavities (Fig. 3). Their plane of division was usually perpendicular to the ventricular surface and part of their surface membrane was exposed to the ventricular cavity. A dividing cell was joined to adjacent cells by junctional complexes. No mitotic cells were seen farther away from the ventricular lining in the preneural period.

◀ **Fig. 16.** At E16 outer branches of radial glial cells of type seen in Fig. 32e terminate in enlarged spherical formations (*EK1*). End-knobs at the pial surface of type seen in Fig. 32f are flattened against basal lamina (*EK2*). The processes are somewhat swollen by the preparative procedure. *BL*, basal lamina. ×13300

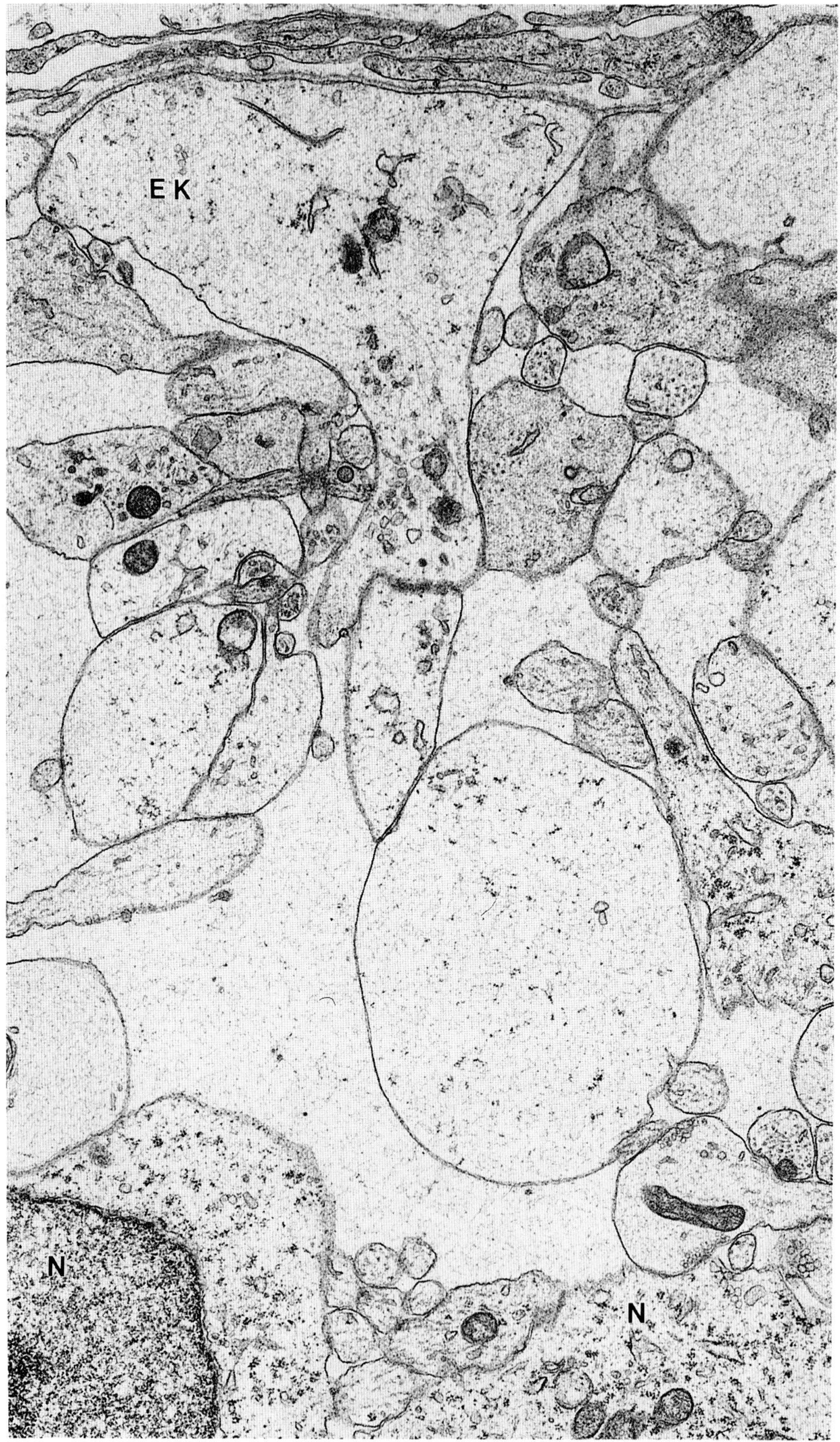

30

4.7 Nerve Cells

Early rounded and oval-shaped nerve cells appeared in the marginal zone on late
E12 or early E13 (Figs. 3, 4, 14). They were situated beneath the pial surface and
were separated from it by the external limiting membrane which was formed by
the end-knobs of the columnar cells (Fig. 15A, B). The cells with oval nuclei and
their processes had an orientation which was parallel to the outer surface and
perpendicular to that of the columnar cells. Later, these and other nerve cells
form layer I in the six-layer cortex (Åström 1967). According to Rickmann and
Wolff (1985) in the rat at E14 some of the horizontal cells in the marginal zone
contact the pial surface and, therefore, are of glial nature.

4.8 Surrounding Tissues

The cerebral hemispheres were enclosed by loose mesenchymal tissue, which in
turn was covered by flat epithelial cells (Fig. 3). Meningeal spaces had not yet
formed. Blood vessels were observed close to the pial surface of the hemispheres
and began to appear within the cerebral wall during E12.

◀ **Fig. 17.** End-knob of an E16 radial glial cell (*EK*) has a wide flat base at the pial surface and a
triangular shape. The cytoplasm is electron lucent and contains SER profiles. *N*, neurons in the
marginal zone. The intercellular spaces are wide. ×13300

5   Area Choroidea

This structure was seen on E11 in a coronal section as a narrow layer of cells
between the two incipient cerebral hemispheres (Fig. 2A). Even at this early
stage of development the cells were shorter than those in the neopallial wall.
During the next two days the area choroidea retained its thickness and width
while the neopallial wall expanded and grew in thickness (Figs. 2B, C, and 18A).
Initially, the area choroidea was contiguous laterally with the hippocampal
anlage. The lateral choroid plexuses started to develop on E13 through invagination at the junction between the area choroidea and the hippocampal anlage
(Fig. 2C). Dorsal to the roof plate and between the two hemispheres was a
region, which contained connective tissue and blood vessels. This is where the
interhemispherical fissure will begin developing at a later stage. As yet the
leptomeninges had not been formed. The telencephalic roof and the choroid
plexus did not contain any blood vessels.

Like the columnar/radial glial cells in the telencephalic wall the roof cells of
the area choroidea were asymmetrical, spanned the entire width of the wall,
were connected near the ventricular surface by junctional complexes, and had a
radial orientation in relation to the ventricular and pial surfaces (Figs. 18B, 28).
The width of a cell was usually largest at the level of the nucleus. The average
length was about 40 µm and the length-width ratio was from 3:1 to 4:1. Thus, the
cells in the telencephalic roof were shorter, wider and more compact than those
in the neopallial wall. Unlike the columnar cells in the telencephalic wall, they

Fig. 18. **A** The area choroidea is demarcated by *arrows*. Above the arrows are the hemispheric ▶
walls, which are seen in this phase micrograph from a fetus at E12. The morphology of the two
regions is different in several respects. The roof cells are shorter, plumper and their enlarged
bulbous ends project into the ventricles. The epithelium contains fat droplets, degenerating
cells and cell fragments, which are seen as dark granules. The ventricular surface of the roof
plate is covered with normal looking cells, with debris, and with necrotic cells. The cells in the
telencephalic wall are longer and more slender. Their inner ends have cilia (indicated by the
serration at the ventricular surface) but no bulbous protrusions. There are no dying cell
fragments and fat droplets in this region, and the surface is not covered with debris. The loose
connective tissue in the interhemispheric fissure contains blood vessels. *LV*, lateral ventricle.
×330. **B** The bulbous protrusions and the terminal bars are well seen in this phase micrograph
of the lamina choroidea in a fetus at E13. The nuclei are situated at different depths of the
lamina. The smaller dots are fat droplets (*arrows*) and the larger dark particles are fragments
of dead cells. Ventricular cavity is in lower part of the picture. ×650. **C** Scanning electron
micrograph of ventricular surface of the telencephalic roof at E13. The bulbous protrusions of
the roof cells look like cobblestones on an old street. The protrusions are largely hemispheric
but the surface of each one has smaller projections and indentations. ×3300

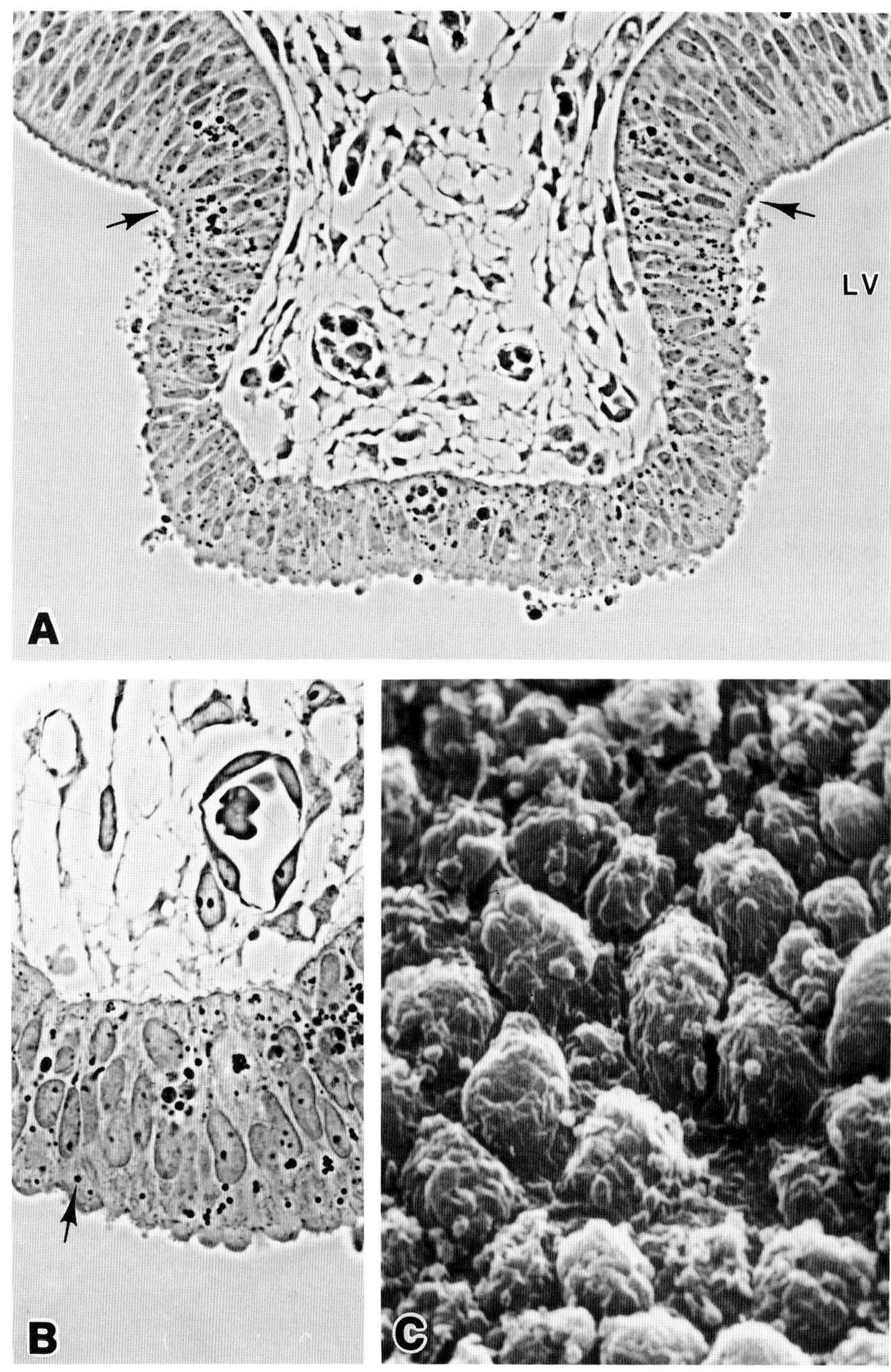

LV
A
B
C

did not increase in length during E11–13. A characteristic feature was that their apical ends extended into the ventricular cavity, forming bulbous protrusions. There were relatively few mitotic cells and, as expected, they were located next to the ventricular surface.

5.1 Perikarya and Nuclei

The nuclei were elongated and had finely dispersed chromatin (Figs. 20, 26). Most of them were located between the junctional zone and the basement membrane, but a few were situated in the bulbous protrusions (Fig. 28). The perinuclear cytoplasm contained clusters of ribosomes but few other organelles.

A characteristic feature of the roof cells was the presence of cytoplasmic inclusions which had the appearance of lipid droplets (Fawcett 1981). When studied in semithin sections with a phase-contrast microscope, they appeared as black dots (Fig. 18B) but were best seen in the electron microscope (Figs. 19, 20, 26). The fat droplets were not seen unless NaCl had been added to the lower ethanol concentrations used for dehydration (Åström and Webster 1990). These lipid inclusions were not surrounded by a membrane. Their size, shape and appearance varied; most of them were rounded and measured from 0.3 to 1.2 μm in diameter. The smallest ones were irregular in shape and their peripheral region was usually less electron-dense than the center. The larger lipid bodies (with a diameter of more than 0.5 μm) were denser and better demarcated. The periphery was often serrated, especially in bodies of medium size (0.4–0.9 μm). The largest bodies (with a diameter more than 0.9 μm) were often spherical or slightly oblong, had a smoother border, and were more electron-dense than the smaller ones.

The lipid bodies were seen in all parts of the roof cells including the bulbous protrusions (Figs. 19, 20) but appeared to be most numerous near the poles of the nuclei in the inner and outer portions of the cells, where they formed groups (Fig. 19A). They did not have any special location in relation to mitochondria, endoplasmic reticulum, or other organelles. These lipid bodies were seen not only in normal roof cells, but also in degenerating, dead, macrophage-like cells, and intraventricular cells. They were also observed in some dividing cells.

5.2 Inner and Outer Processes

The inner and outer processes of the roof cells were as wide as the perinuclear region or somewhat thinner. A Golgi complex occupied a large volume of the

Fig. 19A–C. Rounded fat droplets in roof cells are electron dense due to their affinity for ▶ osmium tetroxide. Many of them have serrated borders. Examination at higher magnification has shown that they are not surrounded by a membrane. Compare these electron micrographs with the phase micrographs in Fig. 18A and B. **A** Outer portions of two E12 roof cells contain 16 and 3 droplets respectively. Intercellular spaces are comparatively wide. ×12500. **B** Inner portions of two E13 roof cells contain groups of fat droplets. Bulbous protrusions are irregular in shape and have many pinocytotic depressions. ×9900. **C** One fat droplet is situated near the ventricular surface at E12. The bulbous protrusions are unusually long. ×13000

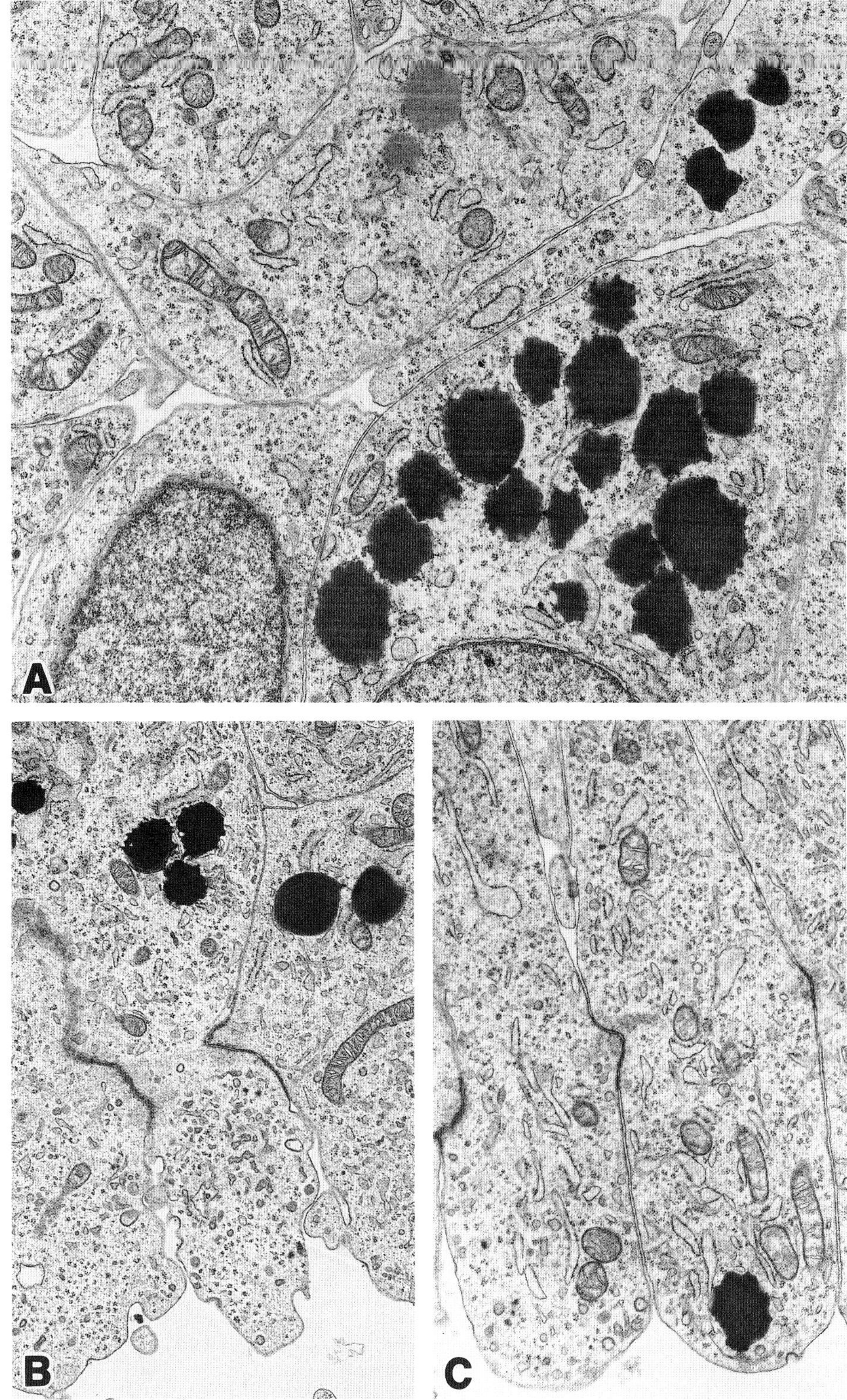

A
B
C

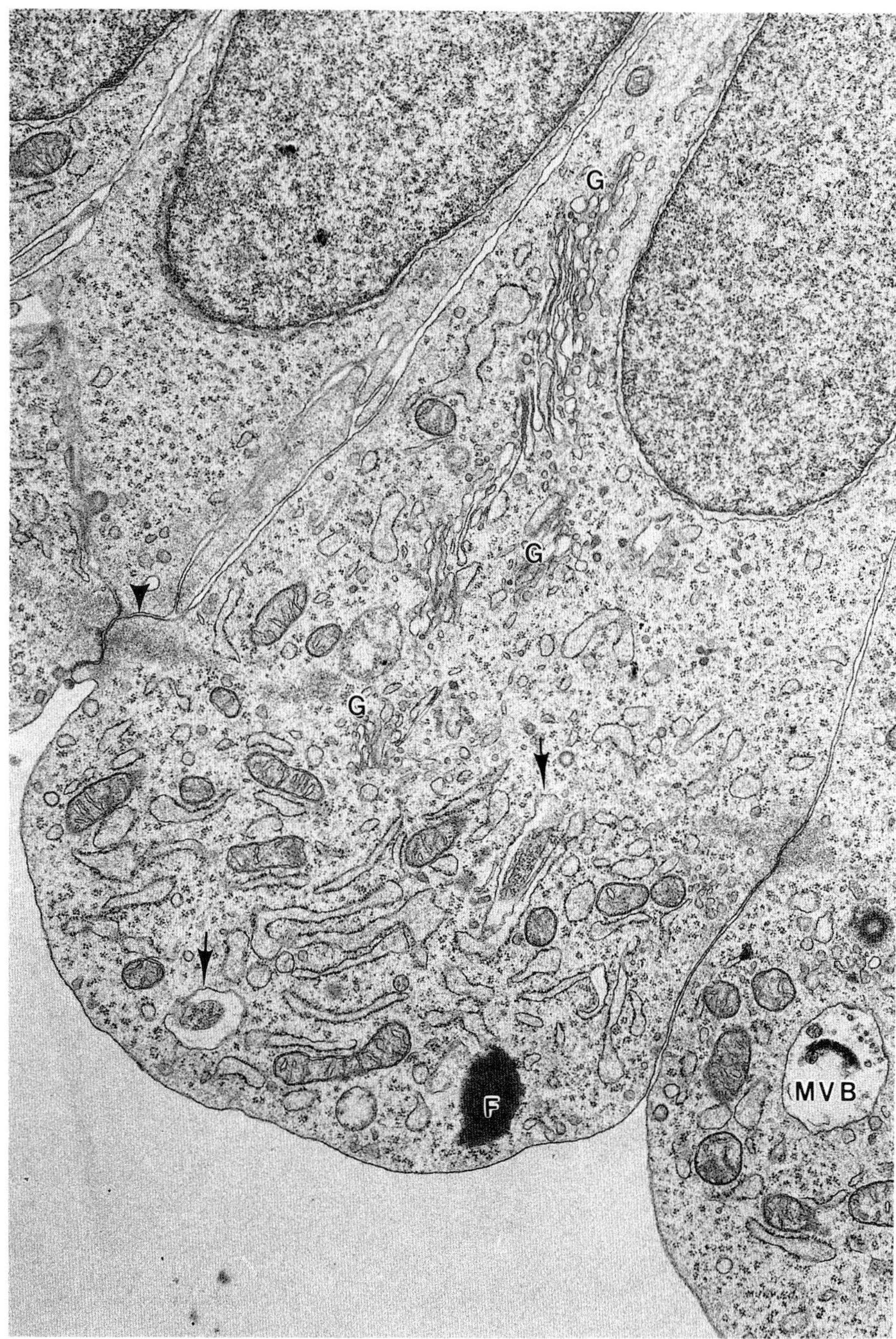

Fig. 20. A Golgi apparatus (*G*) in an E12 roof cell extends from the juxtanuclear region into the inner process and the cell's bulbous protrusion contains one fat droplet (*F*), many mitochondria, and profiles of ER with electron-dense contents. A multivesicular body (*MVB*) contains vesicles and debris. Rosettes of ribosomes are seen everywhere in the cytoplasm. The elongated nuclei contain finely dispersed chromatin. The ventricular cavity is at *lower left*. *Arrow* marks autogenous cytosegresome, and an *arrowhead* marks a junctional complex. ×15000

36

inner process (Fig. 20), where it often extended from the inner pole of the nucleus to the level of the junctional zone or, less commonly, into the bulbous protrusions. The length was up to one-quarter or one-third of the total length of the cell. The cisternae of the Golgi apparatus were in most cases oriented axially with their convex surfaces facing the lateral sides of the cells. A Golgi complex was never seen in the outer processes.

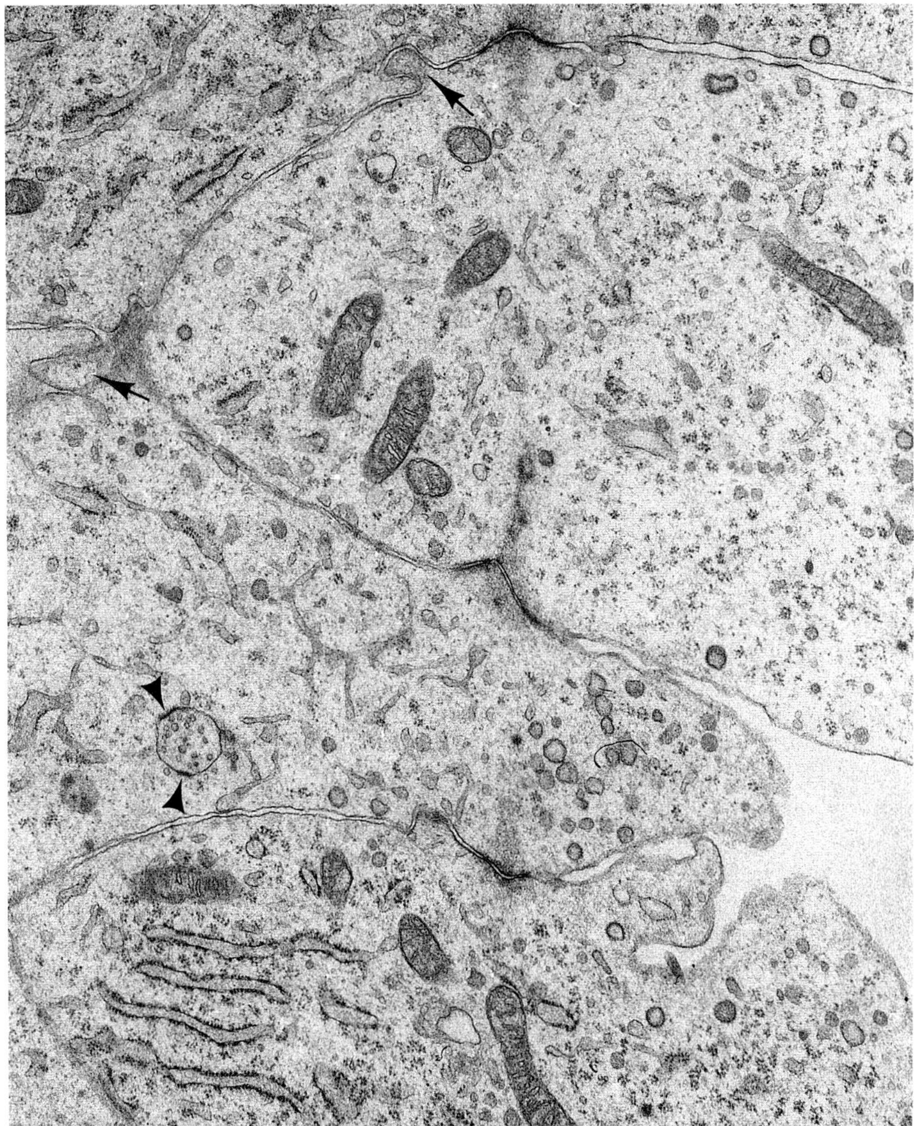

Fig. 21. Three E13 roof cells face the ventricular cavity in the *lower right corner*. The lower cell contains RER, the middle cell contains ER which is mostly smooth, and the upper cell has ribosomal rosettes, mitochondria, and vesicles but no profiles of granular ER. The membrane surrounding a MVB is thickened in two patches (*arrowheads*). *Arrows* point to jigsaw-like interdigitations. ×17000

The roof cells, especially the inner portions, contained the following types of vesicles: (a) Rounded vesicles, with a diameter of 40–80 nm, were seen in or near the Golgi complex (Fig. 20), within and near multivesicular bodies (Fig. 25), and within heterophagic vacuoles. They were empty or contained electron dense material and were probably primary lysosomes which had originated in the Golgi apparatus. (b) Other vesicles near the Golgi apparatus (Fig. 20) and the SER appeared to be profiles of these organelles in cross-section. (c) Multivesicular bodies (MVB) were larger vacuoles, which contained small vesicles, fragments, debris, and floccular material (Figs. 20, 21, 23A, D, and 25). Some of them were probably heterophagic vacuoles which appeared in the course of so-called programmed cell death (Chapter 5.6); others may have contained material from autophagic cytosegregation (Ericsson 1969), a common phenomenon in the roof cells (Figs. 20, 25B).

Mitochondria were seen in all parts of the roof cells but were more numerous in the inner processes (Figs. 20, 21). In phase micrographs they were axially oriented, elongated, thread-like structures, and in electron micrographs, they were round or oval, and sometimes branched.

Profiles of RER of varying width were numerous in the inner parts of the roof cells (Figs. 20, 21) and in some of the bulbous protrusions (Fig. 22); they were less abundant in the outer portions of the cells and scanty in the juxtanuclear regions. However, great variations in distribution were noted within single cells and from cell to cell. The RER was oriented axially or more irregularly. The axial cisternae were especially prominent in the inner processes. Some RER segments were located near mitochondria and were parallel both to them and to stacks of Golgi saccules. RER cisternae always contained electron-dense material. RER profiles had the shape of a Y, T or H, giving the impression that they formed an anastomosing system. Some profiles of ER had only a few ribosomes and segments of SER and RER were found to alternate. Smooth tubules were seen especially in the bulbous protrusions. The total volume of SER was much less than that of RER.

Axially oriented microtubules, single or in small groups (Fig. 23A), were present in all parts of the cells except in the end-feet at the pial surface.

Filaments measuring 5–6 nm were numerous in all parts of the roof cells except for the bulbous protrusions where they were scarce or absent. Most filaments were oriented axially.

5.3 Apical Portions (Bulbous Protrusions)

As previously mentioned, protrusion of the apical ends into the ventricular cavities was a characteristic feature of the roof cells. In scanning electron microscope preparations, the ventricular surface of the telencephalic roof looked like a street paved with cobblestones (Fig. 18C). When sectioned the surface of a protrusion had either a smoothly rounded or a more irregular contour (Figs. 20–24). The longest protrusions expanded below the base in club-like fashion and were so closely packed that very little space remained between opposing cell membranes. Only the bottoms of such cells were freely exposed.

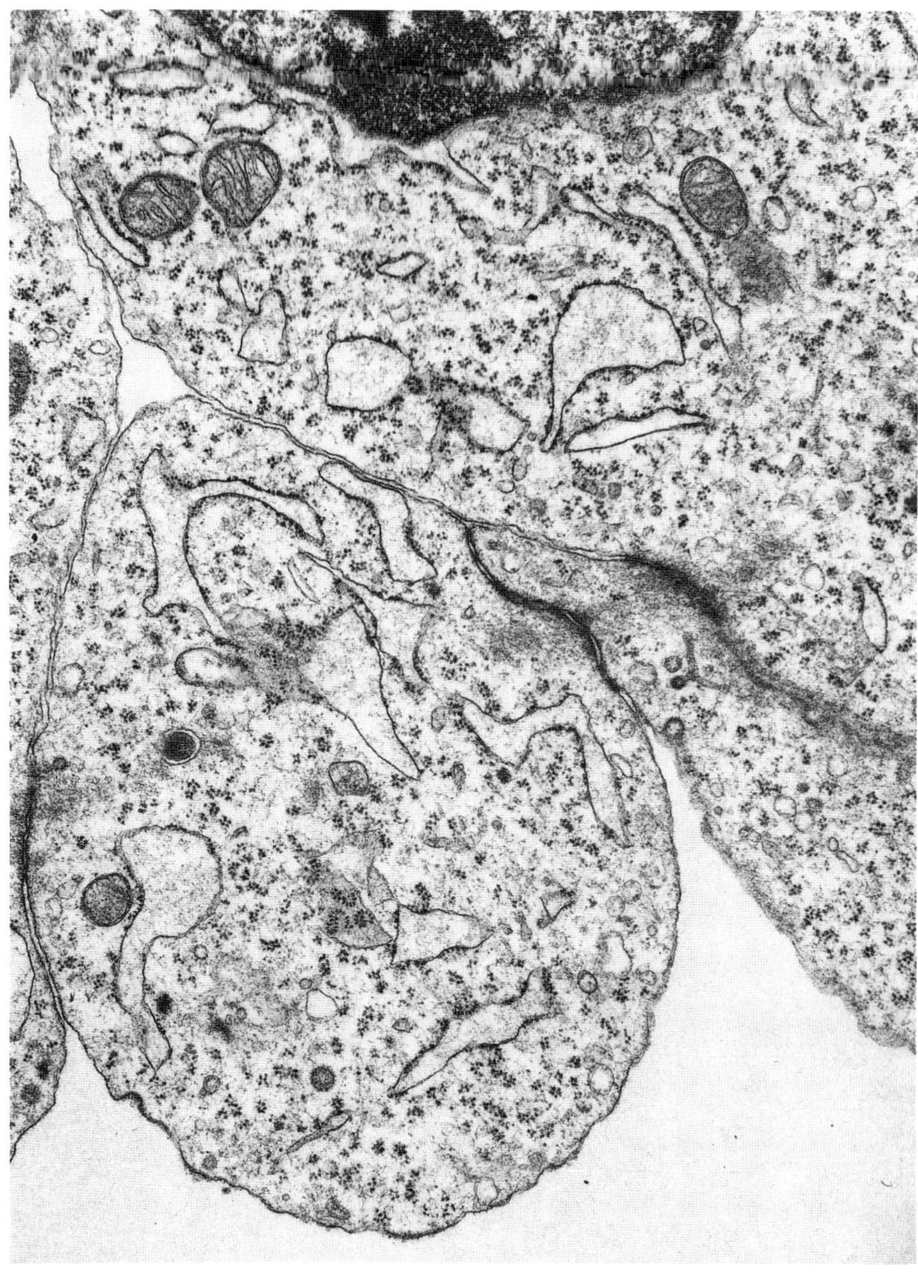

Fig. 22. Inner portion of E13 roof cell and a nearby bulbous protrusion with distended profiles of RER containing flocculent material. ×21000

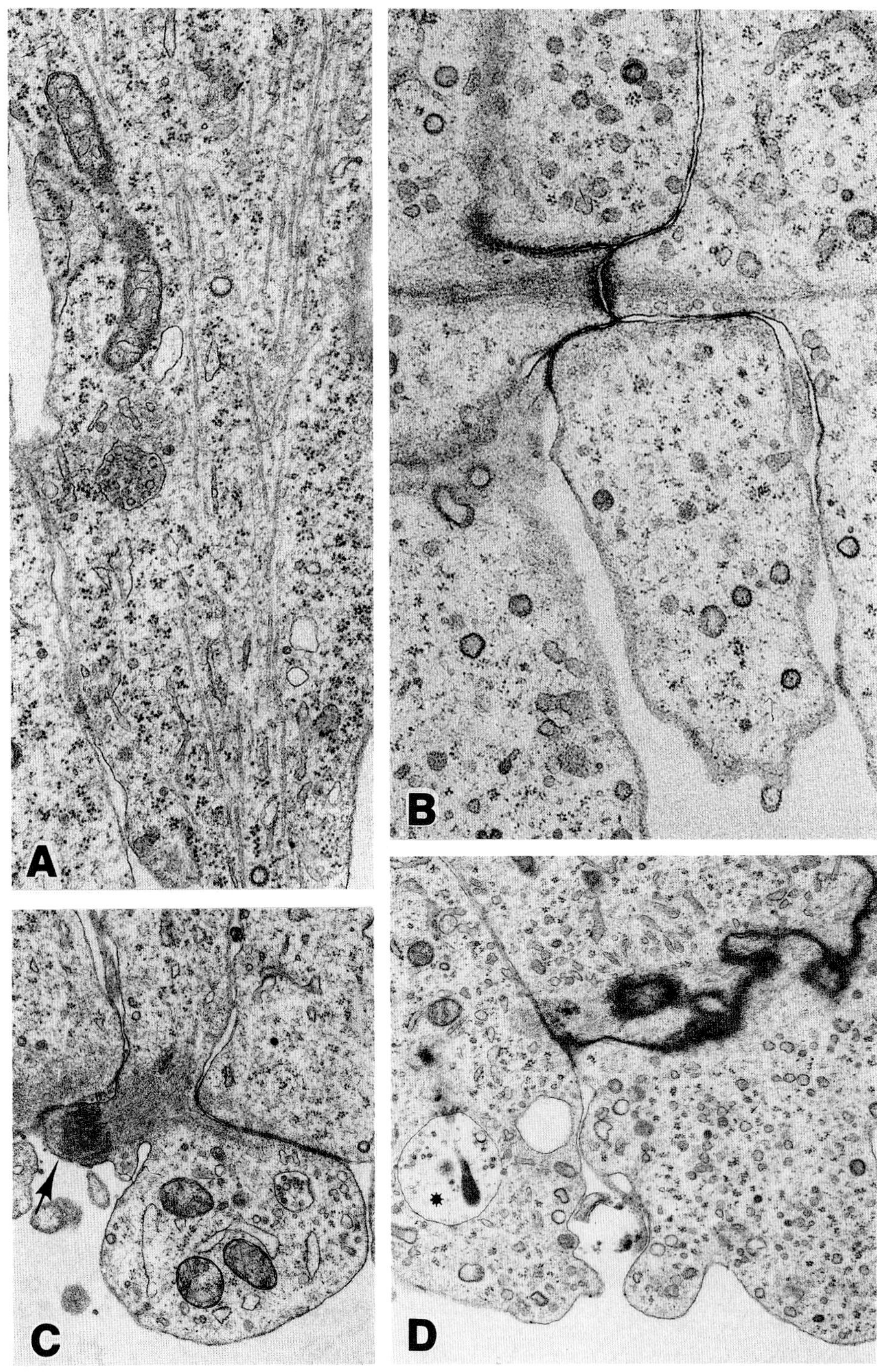

A
B
C
D

Circumferentially arranged microfilaments, attached to thickenings beneath the surface membrane, formed collars at the ventricular ends of roof cells (Fig. 23). This "terminal web" of microfilaments (called terminal bars in the light microscopic literature) defined the boundary between the protrusions and the inner processes of the roof cells.

The surface membrane of protrusions was somewhat thicker than that of the remainder of the cell (about 10 nm versus 7 nm) and had sparse, fine, short filaments projecting into extracellular spaces (Fig. 24). There also were patches of membrane thickening which measured about 0.1–0.3 µm in diameter (Fig. 24C). Filamentous material or small particles or both usually adhered to these areas which also had bristle-like projections on their inner, cytoplasmic surface. These patches were flat or, more commonly, depressed. Coated vesicles were seen in all parts of the cells but were most numerous in the bulbous protrusions and often seemed to be concentrated near the plasmalemma where some of them were connected to the surface by neck-like formations (Figs. 24, A, C, D, F). They had an average diameter of 0.2–0.3 µm, contained electron-dense material and occasionally, small particles. It was noteworthy that the number of coated vesicles did not seem to exceed that of the thickened patches on the surface. Profiles were observed which suggested that vesicles merged with each other (Fig. 24F) or with MVBs (Fig. 25C).

MVBs with a diameter of 0.45–1.3 µm were most numerous in the bulbous protrusions. They appeared as light dots in phase micrographs (Fig. 18B), but were best seen with the electron microscope (Fig. 25). They contained small vesicles and loose amorphous material or appeared to be empty. A few MVBs contained larger particles (Fig. 20). The number of small vesicles within the MVBs varied from a few to so many that the MVB was packed with them (Fig. 25B). Small vesicles were also seen in the cytoplasm near an MVB. The vesicles inside and outside the MVBs had the same size and appearance as those in the Golgi complex. The vesicles were thought to be primary lysosomes.

The limiting membrane of the MVBs was either smooth and homogeneous, or had localized densities with an average diameter of 0.1–0.2 µm (Fig. 25D). Some of these spotlike densities had small bristles which projected into the cytoplasm.

Although the patches, depressions, invaginations, and vesicles described were most common in the bulbous protrusions (for instance, six can be seen in Fig. 24D), a smaller number also occurred elsewhere on or near the surface of roof cells.

Few cilia and centrioles were found in the roof cells. When present, a centriole was always located in the inner portion of the cell, usually closer to the ventricular surface than to the nucleus.

◄ **Fig. 23. A** Inner portion of an E11 roof cell contains axially oriented microtubules, mitochondria, ribosomes, and a MVB. ×25000. **B** Filaments forming "collars" around the "neck" of roof cells are attached to the junctional complexes. E13, ×20500. **C** Formation at *arrow* is typical of bridges which exist between two cells before their separation following mitotic division. E12, ×16500. **D** Interdigitations of E13 cells and junctional complexes are cut tangentially. MVB (*) contains small vesicles and debris. ×12500

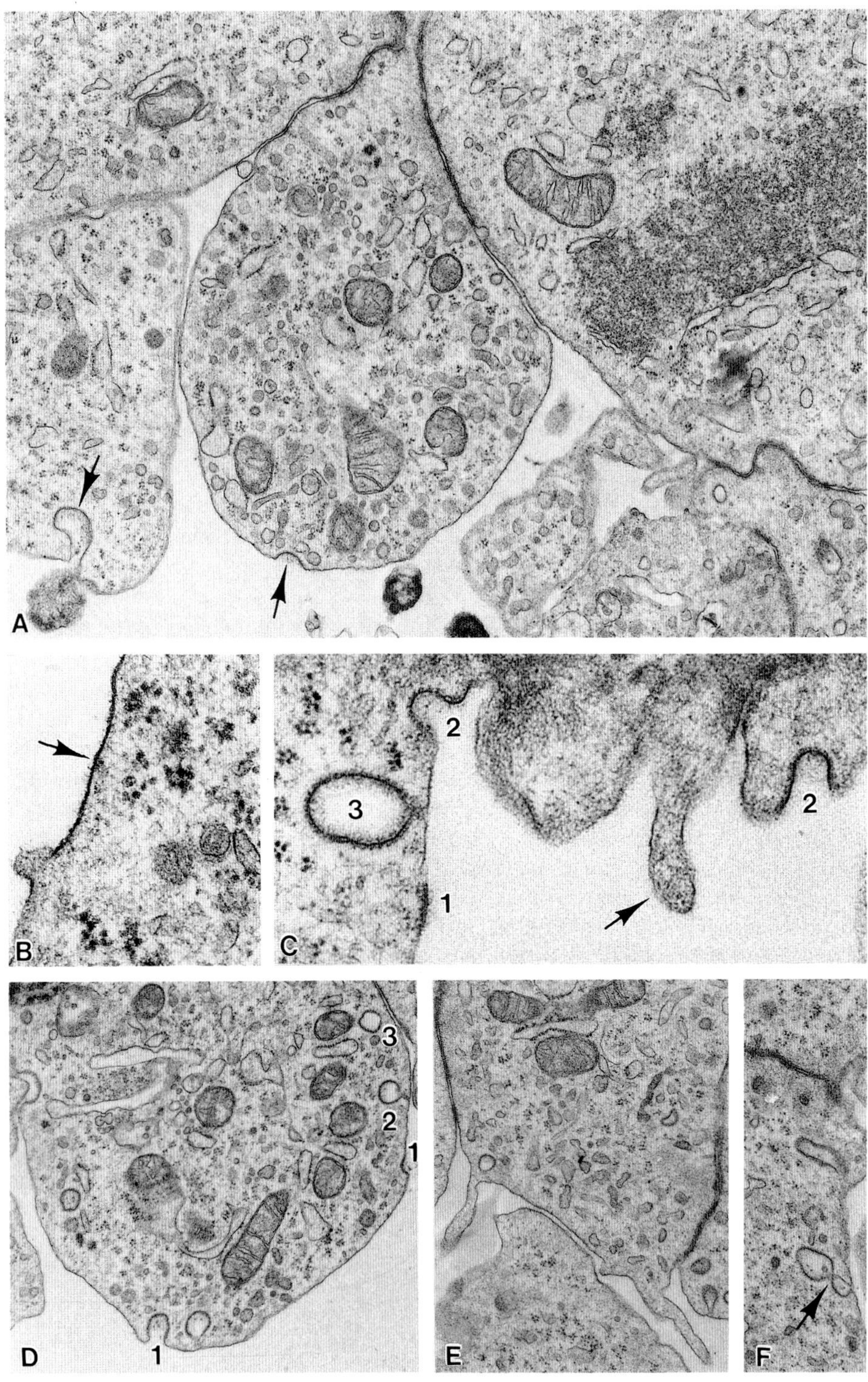
A
B
C
1
2
3
2
D
1
2
3
1
E
F

5.4 Basal Portions

The outer part of a roof cell divided and the branches terminated in interdigitating foot-like processes which rested upon the basal lamina (Fig. 26, see also Fig. 28). Some of these processes were slender, others were as thick as or thicker than the cell bodies from which they originated. Microfilaments formed a dense web peripherally in the end-feet opposite the basal lamina.

5.5 Relations Between Roof Cells

At the level of the terminal bars, the intercellular space was about 5 nm wide and was completely obliterated in spots. Distal to this "neck region" or junctional zone, the space varied usually between 15 and 60 nm, but was occasionally wider, especially in the outer regions.

The lateral sides of the main portion of a roof cell were usually straight and more or less parallel. Occasionally, however, a finger-like process indented its neighboring process, producing a jigsaw-like figure. These formations were seen only at the proximal and distal ends of the cells (Figs. 21, 26B).

The intercellular spaces appeared empty with two exceptions: they contained fragments of necrotic cells and small amounts of fuzzy material where the cells were closely apposed.

5.6 Cell Death

A striking and characteristic feature of the area choroidea of the rat embryo during E11–13 was the presence of degenerating cells, necrotic cells and phagocytes which were characteristic features of programmed cell death (Figs. 18A, B). Fragments of necrotic cells were engulfed and phagocytized by surviving roof cells which contained many heterophagic vacuoles. The ventricular surface of the area choroidea was also covered with cells and debris.

◀ **Fig. 24A–F.** These micrographs illustrate surface membranes, micropinocytosis, and other features in bulbous protrusions facing the ventricular cavity. **A** Indentations (*arrows*) have thickened membranes, bristle-like projections on the cytoplasmic side of the plasmalemma, and usually some flocculent densities on the ventricular side. Mitotic cell in *upper right* part of picture has rounded shape and is still attached to neighboring cells by junctional complexes. E12, ×17000. **B** Thickened surface membrane of bulbous protrusion is covered with fine short filaments (*arrow*). E11, ×53500. **C** Surface membrane has (*1*) flat and (*2*) indented patches. (*3*) Vesicle with thick wall has small opening to ventricle. The membrane of the three formations is thicker than the remaining surface membrane, is bristle-coated on the cytoplasmic side, and has flocculent material on the ventricular or, in the case of *3*, the inner side. Note also the finger-like projection of cytoplasm (*arrow*). E12, ×54000. **D** Bulbous protrusion shows indentation with flocculent debris (*1*) and vesicles with (*2*) and without (*3*) apparent connection to surface. The protrusion also contains ribosomes and mitochondria. E12, ×15500. **E** Bulbous protrusion has fingerlike extensions into ventricle. E13, ×15000. **F** Suspected fusion of two coated vesicles (*arrow*), one of which has an opening to the surface. E12, ×18000

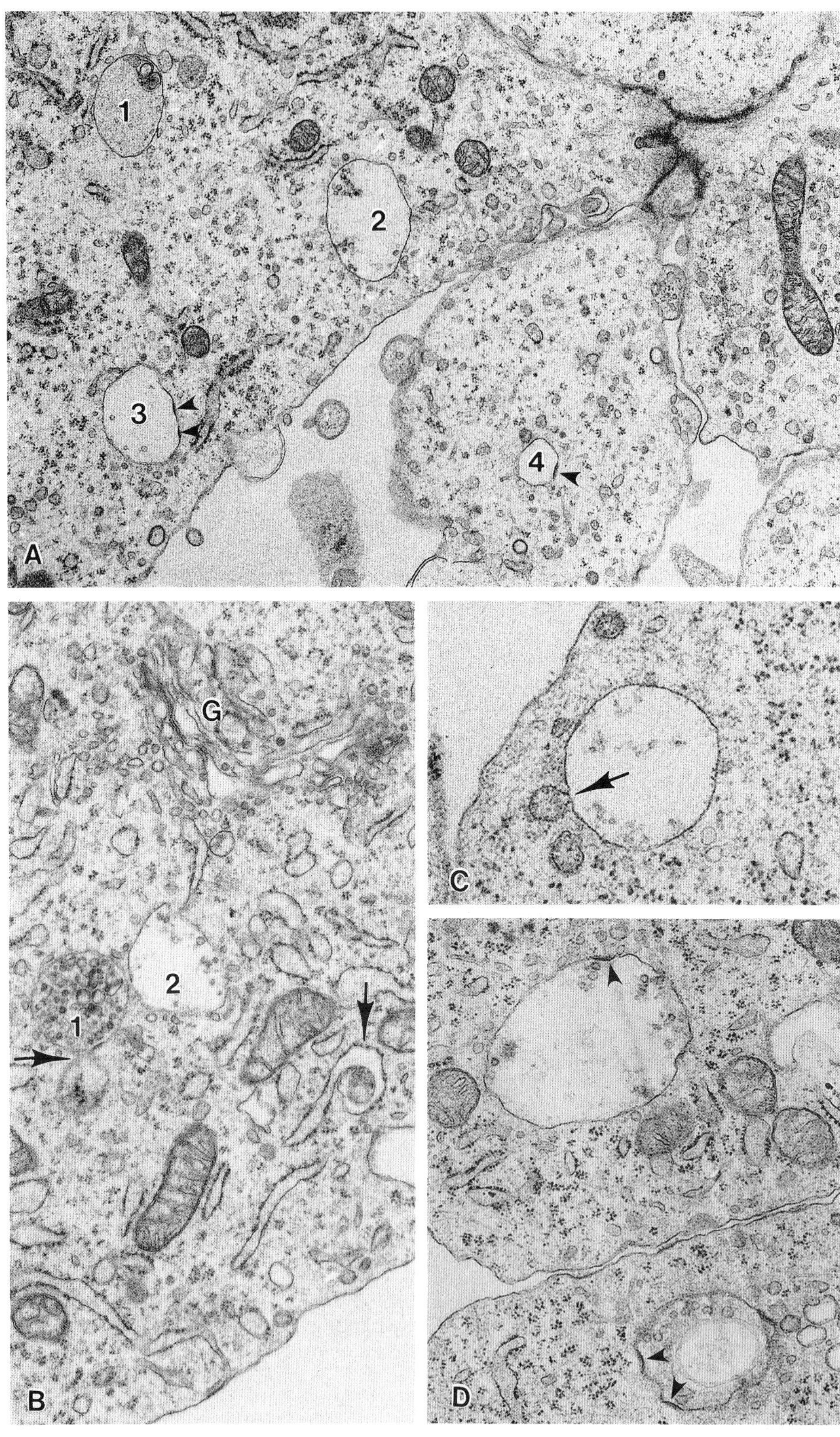

5.7 Comparison Between Columnar Cells in the Area Choroidea and the Telencephalic Wall

In the 11–13-day old fetuses, the cells in the area choroidea and neopallial wall had the following *common* properties (Figs. 27, 28). They were columnar, extended through the entire width of the walls and formed a monolayer which retained some properties of the neuroectoderm. The inner ends had collars of microfilaments, were joined by junctional complexes, and faced the ventricular cavity. The Golgi apparatus was always located in the inner process. The end-feet of the outer processes were reinforced by webs of microfilaments near the basement membrane. Microtubules occupied the entire length of the cells but were not seen in the end-feet.

The cells in the two locations *differed* in the following respects. The *cells of the area choroidea* were short and relatively thick and projected into the ventricular cavity in a bulb-like fashion. They neither elongated nor changed in shape during the E11–13 period. In particular, the outer processes did not ramify and the end-feet at the pial surface did not transform into end-knobs of the type seen in the cells in the neopallial wall. Mitoses were infrequent. The inner processes contained a voluminous Golgi apparatus, many vesicles of various types, mitochondria and much SER. Pinocytotic indentations, vesicles, and MVBs were numerous in the bulbous protrusions. The presence of fat and the occurrence of cytosegregation were characteristic features of the roof cells. In addition, these cells had few cilia. No blood vessels were seen in the area choroidea.

In contrast, *the cells in the telencephalic wall* were longer and thinner and their ventricular ends were blunted and ciliated. They appeared to be metabolically less active than the roof cells since the Golgi apparatus was smaller, vesicles were fewer, pinocytosis was less common, and fat droplets were not seen. Mitoses were common. Cell necrosis and heterophagia did not occur, cytosegregation was rare and the ventricular surface was not covered with necrotic cells and debris. The distal processes of these cells, which rested upon the basal lamina, were more electron lucent because of their lower content of filaments and other organelles. Therefore, they were more difficult to preserve without distortion than corresponding apical ends of the roof cells. The outermost part of the wall (the prospective marginal zone) had no nuclei and did not contain blood vessels.

◀ **Fig. 25A–D.** Micrographs illustrating MVBs. **A** One bulbous protrusion has three MVBs (*1–3*) and another has one (*4*). *1* contains fragments, vesicles, and flocculent material; *2* and *3* contain small vesicles. Similar vesicles are also seen outside the MVBs. *3* and *4* have thickened patches on their limiting membranes (*arrow-heads*). E13, ×12700. **B** A bulbous protrusion contains two MVBs (*1,2*). *1* is packed with small vesicles; *2* has only a few of them. Small vesicles are also seen outside these MVBs and close to the nearby Golgiapparatus (*G*). A larger vesicle with debris appears to be merging with *1* (*left arrow*). The *right arrow* shows an area of cytosegregation. E12, ×18500. **C** A coated vesicle appears to be merging with a MVB (*arrow*). Two other coated vesicles are nearby. E11, ×16000. **D** Two MVBs have patches in their limiting membranes (*arrow-heads*). E11, ×34000

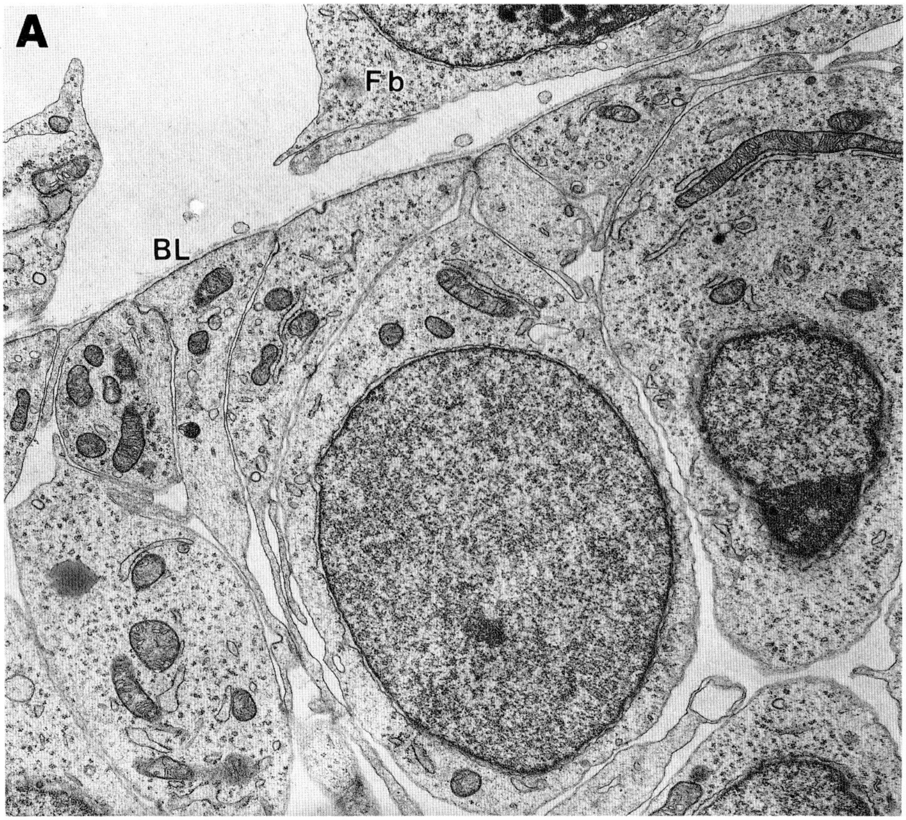

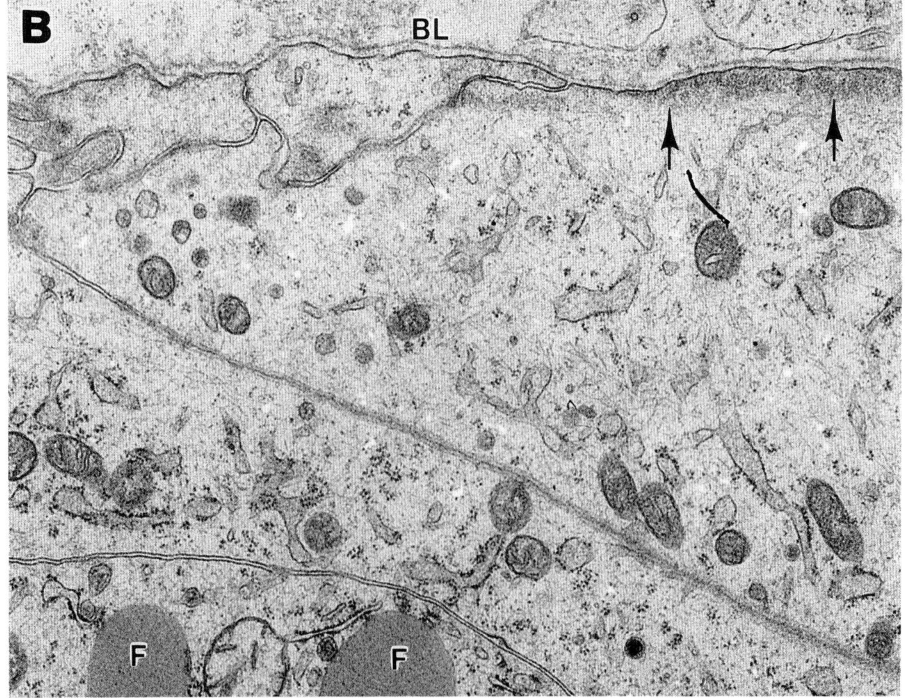

A
Fb
BL
B
BL
F
F

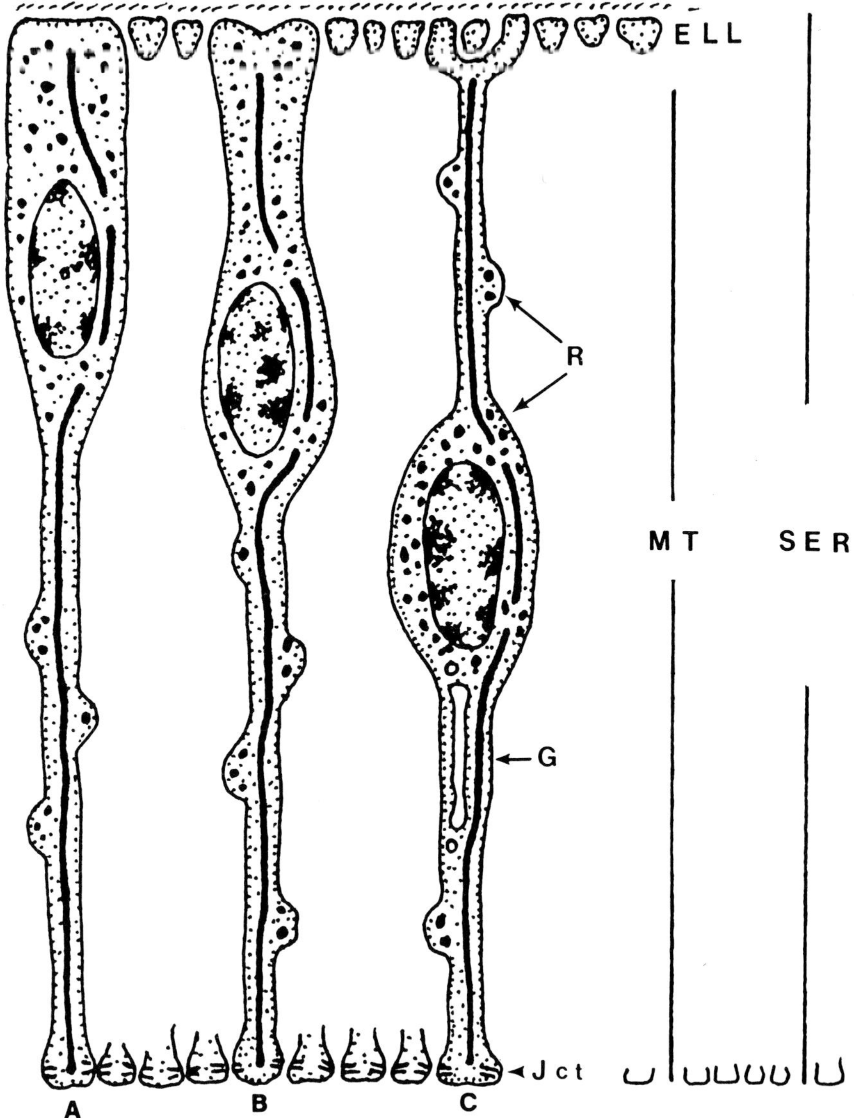

Fig. 27. Diagram showing essential features of three neopallial wall bipolar cells (dimensions not to scale). *A*, *B*, and *C* show the changing distribution of cytoplasm in cells as the outer process increases in length. *ELL*, external limiting layer; *G*, Golgi apparatus; *R*, ribosomes; *MT*, microtubules; *SER*, smooth endoplasmic reticulum in cell *C*; *Jct*, junctional complexes joining ventricular ends of inner processes

◀ **Fig. 26A,B.** Outer portion of roof cells. **A** Processes of varying shape reach or are directed towards the outer surface of the lamina choroidea which is covered by a basal lamina (*BL*). Slender processes enlarge peripherally and form end-feet which contain ribosomal rosettes. There are fewer vesicles and tubules of SER (probably also mitochondria) than in the inner portion of the cells. There is no Golgi apparatus in the outer processes. Intercellular spaces are in places comparatively large. *Fb*, fibroblast. E12, ×9500. **B** Filaments form a web beneath the basal lamina (*BL*) and in adjacent areas (*arrows*). Note the interdigitation of cellular processes. *F*, fat. E13, ×18500

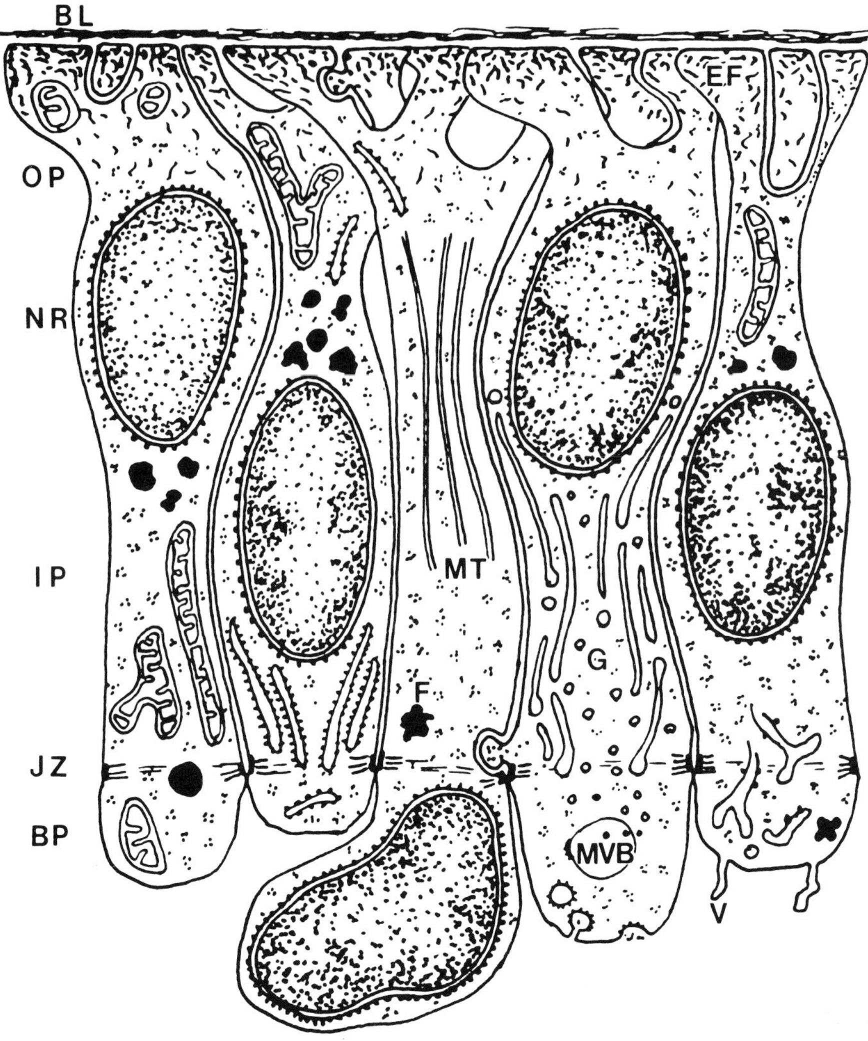

Fig. 28. Diagram summarizing the salient features of cells in area choroidea (dimensions not to scale). The cells are polarized; beginning at the ventricular surface five parts can be recognized: bulbous portrusion (*BP*), junctional zone (*JZ*), inner process (*IP*), nuclear region (*NR*), and outer process (*OP*). One cell has its nucleus in its bulbous protrusion. A large Golgi apparatus in the inner process (*G*), numerous ribosomes, profiles of RER and SER, fat droplets (*F*), multivesicular bodies (*MVB*), vesicles, and granules reflect higher metabolic activity and possibly also more secretory and absorptive functions in these cells than exist in the columnar cells that form the neopallial wall (Fig. 27). The roof cells also contain microtubules (*MT*), and filaments. *BL*, basal lamina. *EF*, end foot

6   Discussion: Neopallial Wall

The methods which were used in this study (Åström and Webster 1990) have allowed us to examine whole brains of fetal rats in an early period of development (E11–13) and to use light and electron microscopy for studies of well-preserved cells within selected areas of the pallial wall. During this time period the selected regions (the neopallial wall and the area choroidea) have the same basic structure, which is that of a pseudostratified neuroepithelium. Like authors of previous studies, we found that the neuroepithelial cells are radially oriented and have an asymmetrical shape (His 1889; Sauer 1935b). They are joined by junctional complexes near the ventricles (Duncan 1957), they extend through the wall to the pial surface where the outer ends are covered by a basal lamina, and they have cytoskeletons composed of axial microtubules and microfilaments (Lyser 1964, 1968; Herman and Kauffman 1966). We have also confirmed previous observations on the fine anatomy of the columnar/radial glial cells in the telencephalic wall of mice (Meller et al. 1966; Hinds and Ruffett 1971; Choi 1988), in rats (Matsuyama et al. 1973; Peters and Feldman 1973; Rickmann and Wolff 1985), and in rabbits (Stensaas and Stensaas 1968). Earlier studies are concerned mainly with the radial glial cells which appear during the period of neuronal migration. This project is different in that it deals only with the columnar cells which are seen in the preneuronal period. Of particular interest are changes in the shape of the columnar cells, the transformation of internal structure which takes place when a spherical daughter cell becomes elongated after mitosis, and some previously less well described cytological features, especially the appearance of the plasmalemma and membranous organelles, and the shape, size and location of the Golgi apparatus.

This study also includes the fine structure of the area choroidea, which has not been described previously. We have shown that the columnar cells in the neopallial wall and area choroidea have the same basic structure, but that they differ with respect to individual features such as size, shape, and internal structure. These variations reflect cellular functions, which are different within the two areas.

Our observations raise a number of questions concerning the development of the brain in a mammalian animal like the rat. How do the columnar cells change during the intermitotic periods? What forces are involved? How are the cerebral hemispheres in the mammalian brain shaped? What are the functions of the cells in the neopallial wall and telencephalic roof during the early period of development? These and other questions will be discussed here.

Our interpretations have a limited scope since they are based on static images and not on direct observation of living specimens. Even so, we think our data and conclusions will help other neuroscientists in this field by providing a basis for comparisons with cultured cells and explants, with cells seen in lineage studies, and with microscopic observations of living animals in which dynamic events in the CNS can be seen directly.

6.1 Nature of Columnar Cells

Our observations and those of others (Choi and Lapham 1978; Schmechel and Rakic 1979; Varon and Somjen 1979) strongly suggest that the columnar/radial glial cells are the first and main constituents of the early CNS.

Interphase and mitotic cells in the preneural phase (E11–12) of the telencephalic wall development form a monolayer which only contains neuro-epithelial cells and a few blood vessels. The columnar cells, also known as spongioblastic, epithelial, bipolar, and radial cells, are homologues of the cuboidal cells that form the neural tube and neural plate.

After E13 the simple structure of the telencephalic wall is modified by the appearance of morphologically different cells, the neurons, which migrate outwards and settle in the outer part of the wall. At this stage of development the columnar cells are longer, have thinner processes and are called radial glial cells (we restrict the use of this name to those columnar cells which are seen during the period of neuronal migration). Hence, the wall's homogeneous columnar cell structure is gradually obscured by the appearance of more and more neurons and radial glial cells, and will disappear completely when the neurons have developed processes, and when all radial glial cells have been replaced by mature astrocytes.

Radial glial cells differ from columnar cells in several respects. They contain intermediate filaments that can frequently be stained by antiserum raised against glial fibrillary acidic protein (GFAP) (Valentine et al. 1983; Choi 1988). They contain glycogen granules (Gadisseux and Evrard 1985) and have terminal knobs at the pial surface which are electron lucent and PAS positive (Peters and Feldman 1973). Compared to columnar cells, radial glial cells also have longer and thinner processes.

On the other hand, light and electron microscopic studies have shown that radial glial cells and columnar cells in the early telencephalic wall and the cuboidal cells in the neural tube *share* the following morphological features: They are bipolar, asymmetrical, radially oriented, and span the entire width of the wall. They terminate freely at the pial surface and have cytoskeletons that include axially oriented microtubules and circumferential arrays of micro-filaments beneath their junctional complexes. A major purpose of this paper is to stress the basic similarities and to discuss the morphogenesis and function of these closely related cells. Two conclusions of particular significance emerge from this review: (a) These cells are attached to each other and together form an epithelial sheet which is similar to that in the neural tube and plate and indeed to that in any sheet of epithelial columnar cells in other organs of the body. (b) The cells are morphologically polarized. We will see that these conclusions are important for an understanding of early CNS development, morphogenesis, and function.

50

In order to simplify the description, we have used different names for the cells at three developmental stages. Cuboidal cells seem to evolve into columnar cells and columnar cells into radial glial cells. However, these cells are dividing continuously and, therefore, a family of cells exists in which morphological changes occur in small steps from one generation to the next. For instance, radial length increases greatly during the transition from columnar cells in the neural tube to radial glial cells of the telencephalic wall. The outer process (between the nucleus and the pial surface), which is short and plump during early stages of development (Fig. 27), becomes longer, more slender, and forms branches, which terminate in knobs at the pial surface. The internal structure is also modified: the Golgi apparatus is elongated (but always located in the internal process), the terminal knobs become more electron lucent and glycogen granules appear throughout the cytoplasm. The cytoskeletal changes are especially noteworthy; vimentin, an intermediate filament constituent, appears early and is later replaced by GFAP.

Cajal (1909) concluded that the epithelial cells emanate from "spongioblasts" and develop into astroglia and ependymal cells. This idea has been confirmed by two sets of observation. First, the radial glial cells in some species contain GFAP (Choi and Lapham 1978; Levitt and Rakic 1980), an intermediate filament protein which is thought to be specific for astrocytes. Secondly, some or all radial glial cells develop eventually into astrocytes (Cajal 1909; Schmechel and Rakic 1979). Thus, radial glial cells are currently thought to be members of the astrocyte family (Fedoroff 1986). In conclusion, the epithelial cells in the neural tube and early telencephalic wall are columnar cells which during development gradually acquire features of astroglia.

Earlier neuroanatomists (Lorente de Nó 1933) claimed that glial cells appear and develop *after* neurons. This concept is only partly correct. It is true that neuroglial cells of adult type appear *after* the nerve cells but radial glial cells and their morphologically recognizable precursors clearly appear *before* any nerve cells (Choi and Lapham 1978). Evidence from many studies shows that newly formed neurons could not move to predetermined locations and develop processes with complex connections over long distances without a preexisting glial framework to guide neuronal migration and process growth.

6.2 Columnar Cell Mitosis and Radial Growth

As first shown by His (1889), the wall of the early neural tube consists of epithelial cells and mitotic cells. As the wall grows in thickness, the nonmitotic cells become *columnar* in shape. He believed that the columnar cells were supportive and spongioblastic, and that the mitotic cells gave rise to neuroblasts only. The origin of the spongioblastic cells was unclear.

This view has been challenged and a fundamental question in neuroembryology for many years has been: Do the neuroepithelium and neural tube contain one or several cell-lines? The prevailing view now is that mitotic and epithelial cells form a morphologically uniform population of germinal cells (Sauer 1935a; Sauer and Chittenden, 1959; Sauer and Walker 1959; Sidman et al. 1959; Fujita 1963; Watterson 1965). It should be emphasized that this concept refers only to morphological homogeneity. Although all cells in the early neural wall seem to go through these changes it cannot be excluded that different cell lines exist in the neural tube or, indeed, in the neuroepithelium.

Preparing for division, a bipolar cell (Fig. 29a) loses its "grip" at the outer, pial surface and "collapses" (b). The nucleus appears to migrate towards the lumen, the cytoplasm contracts around the nucleus, and the cell assumes a spherical shape at the ventricular lining as it enters the mitotic phase (c). After division the daughter cells (d) extend radially through growth of their processes (e, f). The nuclei and cytoplasm "migrate" outwards and the outer processes reach the pial surface again (g). This cycle of changes, which is repeated many times, involves a to and fro movement of nuclei within the neopallial wall, which increases in width as the wall increases in thickness. Although the change in cell shape and the displacement of nuclei during interphase are spectacular, the dynamics of division are essentially the same as they are in other epithelial organs, e.g. the thyroid gland (Zeligs and Wollman 1979; see also Sauer 1935a). Of central importance is that mitosis occurs near the central lumen where parent as well as daughter cells are joined to their neighbors by junctional complexes. Therefore, the neopallial wall is preserved as a continuous monolayer and CSF cannot escape to the intercellular spaces either during or after cytokinesis. Moreover, the daughter cells retain the same relative position that their parent cells occupied in the mosaic of columnar cells. The premitotic collapse of a parent cell, the postmitotic extension of the daughter cells, and the "elevator movements" of nuclei (Fujita 1963) are secondary events and related to the fact that the columnar/radial glial cells are not only germinal cells but also serve during interphase as building blocks of the neopallial wall which is growing thicker.

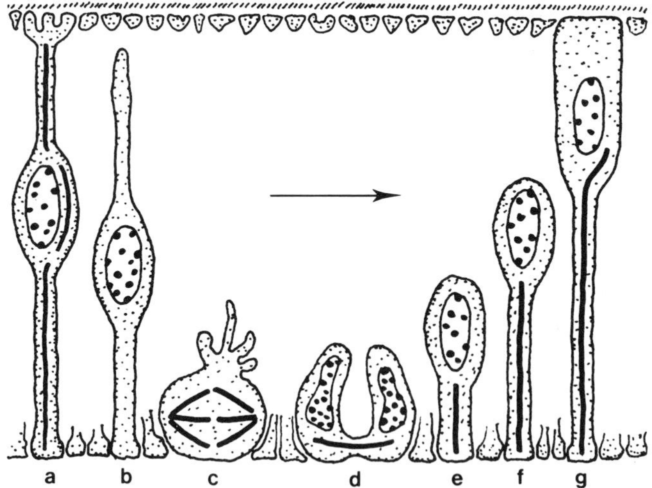

Fig. 29. Diagram suggesting sequence of changes in a columnar cell during early development of the telencephalic wall (cf Sauer 1935a, Fig. 8): A columnar cell (*a*), collapses due to disassociation of microtubules (*b*), and divides at the ventricular surface (*c, d*). A daughter cell regains original shape and length (*e→f*) and later increases in length (*g*)

The development of polarity is inherent in the structure of the columnar epithelium (Handler 1989). The following anatomical features ensure that the daughter cells maintain the polarity of the parent cell: (a) a bipolar cell keeps its attachment to the neighbors as it collapses and divides at the ventricular surface; (b) the plane of cleavage is perpendicular to the apical aspect of the parent cell; (c) new tight junctions are formed between the daughter cells basal to the cleavage bridge.

Interest has been focused on the "migration" of nuclei. Hinds and Ruffett (1971) for instance, suggested that contraction of filaments in a columnar cell during prophase causes movement of its nucleus inward. We propose, instead, that microfilaments play a secondary role and that the depolymerization of tubulin produces a collapse of the bipolar cells in prophase (cf Sauer 1936). This idea is supported by the following evidence: (a) Microtubules play a role in the elongation and maintenance of cell shape (Bayers and Porter 1964) including the shape of cells in the neural plate (Waddington and Perry 1966; Karfunkel 1971; Burnside 1971). (b) Microtubules are formed by the polymerization of alpha and beta tubulin; their formation and dissociation depend on the functional needs of the living cell (Burnside 1975).

In conclusion, we suggest the following developmental sequence. The dissociation of microtubules triggers the collapse of a cell (Fig. 29b), which is pulled by arrayed microfilaments to its anchor point at the ventricular surface where it assumes a spherical shape. The contraction and reduction of the cell surface results in an excess of membrane which will surround the filopodia at the outer end of the dividing cell (Seymour and Berry 1975; see also Fig. 11B in this paper). Dissociated tubulin is used to form microtubules in the mitotic spindle (Fig. 29c). Elongation of the daughter cells is then associated with assembly of axial microtubules from tubulin after the mitotic spindle has been dissociated (e → g). It follows from the above that the premitotic transformation of a cell is a passive process, whereas in the postmitotic phase the daughter cells develop into new bipolar cells through the growth of inner and outer processes. The microtubules alternate between assembled and disassembled states in this sequence of changes. They are dissociated in the premitotic phase (a, b), rebuilt as spindle fibers during mitosis (c, d), dissociated in the daughter cells, and reassembled as the new bipolar cells elongate (e → g) (new tubulin subunits are also added in the latter).

The transformation of a daughter cell into a bipolar element can be conceived of in several different ways. Four alternatives will be discussed here.

1. In the preneuronal period a daughter cell, anchored at the lumen, develops a process between the nucleus and the ventricle (Fig. 30) which "lifts" the nucleus and the surrounding cytoplasm (here called perikaryon) until the outer aspect is flattened towards the inside of the basement membrane (Fig. $30a_1$). Then an outer process develops and "lifts" the basement membrane while growing (Fig. $30a_2$ and a_3). It follows from this hypothesis, that the inner process stops growing once the nuclear region reaches the pial surface, that it retains a constant length while the outer process continues to grow (until the cell again collapses in the next cycle), and that the nuclei of more recently formed cells are situated farther away from the ventricle than those of "older" cells (cf Fig. 30 a_1, b_2 and c_3).

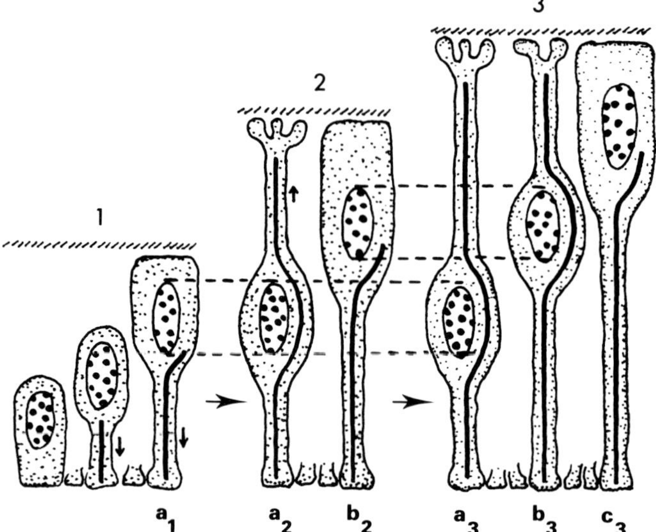

Fig. 30. Suggested scheme for postmitotic development of columnar cells at three stages of development during the preneuronal period according to the first hypothesis. *1*, The inner process of a postmitotic cell grows until the perikaryon touches the pial surface (*a₁*). *2*, A cell similar to a₁ starts to develop an outer process (*a₂*). An a₁-like cell with an elongating inner process reaches the surface (*b₂*). *3*, The outer processes of a₂-like cells continue to grow while the perikaryon remains fixed with relation to the ventricular surface (*a₃*); b₂-like cells develop an outer process, and become *b₃* cells while their perikarya remain in place. Further inner process elongation in a₁-like and b₂-like cells produces c₃ cells

In the period of neuronal migration the cell bodies of the columnar cells are prevented from reaching the pial surface and are stopped first below the layer of neurons in the cortex (Fig. 32b) and later in development below the growing white matter (the intermediate zone) and subventricular zone (Fig. 32c). The outer processes of the bipolar cells will proceed to the pial surface as described above.

2. The nucleus and the main part of the cytoplasm remains at the ventricular surface while the outer process grows until it reaches the pial surface (Fig. 31 II). The nucleus with cytoplasm then migrates through the outer process.

3. The outer and inner processes grow simultaneously (Fig. 31 IV). If both processes grew at the same rate, the nuclear regions would always be situated at the same level and at an equal distance from the ventricular and pial surfaces. To account for an even distribution of nuclei in the ventricular zone the hypothesis as stated would have to be amended to include the assumption that growth rates differ.

4. The inner process first lifts the nuclear region to a position which is somewhere in the middle of the ventricular zone (Fig. 31 III). The outer process then grows and eventually reaches the pial surface.

The first hypothesis (Fig. 30) seems most compatible with known facts: (a) It accounts for the existence of cell type A in Fig. 27 which the other hypotheses do

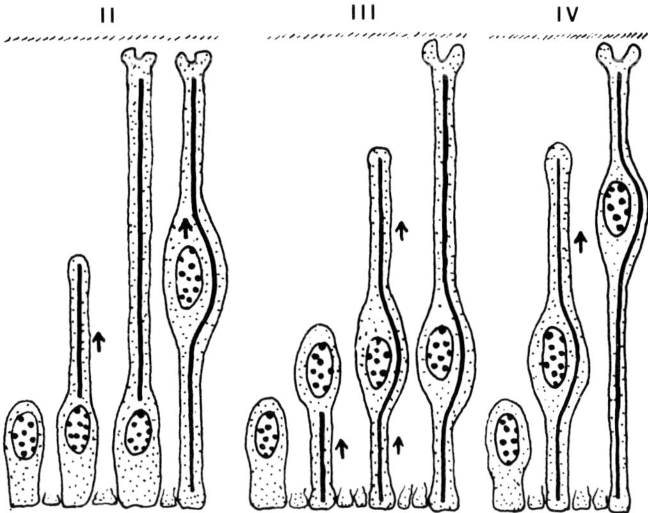

Fig. 31. Illustrates alternative hypotheses to that shown in Fig. 30. (see text)

not unless one assumes that the nucleus and cytoplasm can migrate through the process. (b) It is also compatible with the fact that cells of this type are most common in the early stages of development and are not seen after E14. (c) It explains the even distribution of nuclei throughout the ventricular zone better than the other hypotheses. (d) One does not need to assume that the nucleus migrates. It seems unlikely that a nucleus and surrounding cytoplasm with a combined width of at least 6 µm can migrate through a process which has a diameter of 2 µm or less.

A few more remarks will be made in relation to Fig. 30 and the sequence it illustrates. The word "lifts" gives the impression that the inner process pushes the perikaryon outwards as it grows. However, in reality the process probably grows in the opposite direction, that is away from the nucleus (Fig. 30) as is the case when a nerve cell forms an axon. In the latter case the cell body remains in a fixed position while the tip of the growing axon moves away from the cell body. In the case of the columnar cell, since the tip of the process is anchored at the ventricular lining, the cell body moves away from the ventricle while the inner process elongates. This movement ceases when the cell body reaches the basement membrane (Fig. 30 a_1, b_2 and c_3). The growth of the inner process is centrifugal in the sense that cytoplasm flows along the microtubules from the central regions to the apical ends where tubulin polymerization subunits are added to the peripheral tips of axial microtubules. The axially oriented channels of SER may also have a role in this transport.

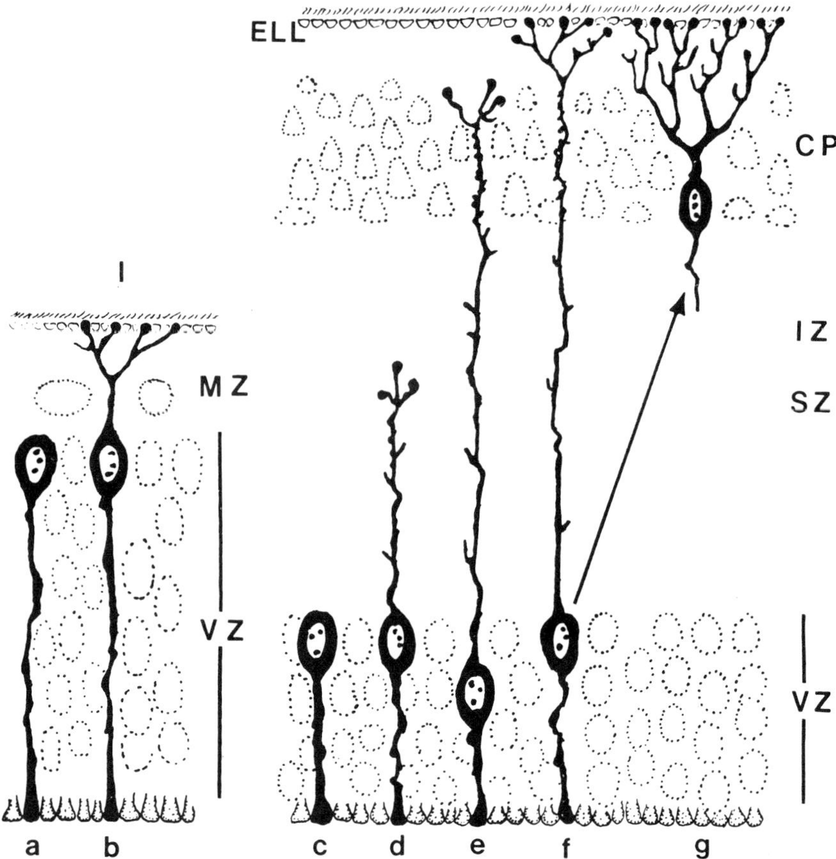

Fig. 32. Diagram showing postmitotic development of columnar and radial glial cells during the period of neuronal migration. It is based on observations by several authors in different species. Columnar cells which have inner but no outer processes (*a, c* in the figure) have been described by Cajal (1909, Fig. 257), Stensaas and Stensaas (1968, Figs. 2, 8), Hinds and Ruffett (1971, Fig. 33). Columnar/radial glial cells in which the outer processes terminate within the wall (*d, e,*) and at the pial surface (*f*) were illustrated by Åström (1967, Figs. 4A, 4B, 8). Superficially located astrocyte (*g*) (Åström 1967, Figs. 15, 18B, 19; see also Eckenhoff and Rakic 1984, Fig. 16). (*b*) is hypothetical. *I*, In the early phase of neuronal migration the inner process of a postmitotic cell grows until the perikaryon touches the superficial layer of neurons (*a*). It then develops an outer process which reaches the surface (*b*). *II*, In the late stage of neuronal migration the inner process of a postmitotic cell grows until the perikaryon reaches the intermediate/subventricular zone (*c*). It then sends an outer process which grows until its terminal, ramified ends reach the pial surface (*d, e, f*). After the period of neuronal migration some (perhaps all) radial glial cells are transformed into astrocytes (*f → g*)

6.3 Cytogenesis of Columnar Cells

After cell division a program is activated which directs the transformation of the rounded daughter cells into polarized columnar cells of epithelial type. This transformation involves *inter alia* elongation of cells and redistribution of cytoplasm. Polysomes and profiles of granular endoplasmic reticulum remain in the perinuclear region (perikaryon) but smaller portions of cytoplasm with such constituents occur as varicosities along the length of the inner and outer processes. The Golgi apparatus is elongated and always located in the inner process. The elongation is associated with assembly of microtubules (Burnside 1971; Porter 1973). Contraction of microfilaments may also contribute to the redistribution of cytoplasm. Furthermore, the cells become aligned with their neighbors and their peripheral ends have a directional movement (Trinkaus 1982, 1984) towards the pial surface.

The mechanisms for directional movement, which have received little attention by neuroembryologists, are less well understood. The situation is simple in an early phase of development, when all cells are relatively short and the telencephalic wall is a monolayer. After mitosis the daughter cells are forced to grow in a centripetal direction since they are anchored at the ventricular surface (which is inflated by CSF pressure) and to move in channels, which are formed by adjacent, fully developed, similarly oriented columnar cells. The situation is more complex at later stages of development when the telencephalic wall is thicker and has a more complicated structure.

It has been postulated that a major function of the radial glial cells in the telencephalon is the guidance of young neurons to their final destinations in the cerebral cortex (Rakic 1971). This hypothesis implies that the radial fibers provide direct well oriented connections between the inner and outer regions of the telencephalic wall and that the spatial relations between the neuronal precursor cells at the ventricular lining on one hand and between their progeny at the pial surface on the other are maintained in a map-like fashion (Rakic 1978, 1982). Consequently, one can conceive of the pial surface as a mosaic of target zones. The outer process of a developing postmitotic daughter cell must seek out, recognize, and connect to its specific target zone. Moreover, in order to do so it must travel over relatively long distances and traverse increasingly complicated networks of processes. This raises an interesting question which has received little attention. Granted that the radial glial cells after mitosis extend their outer process to the pial surface and that these processes guide the young neurons towards predetermined destinations in the cortical plate, what guides the radial fibers towards *their* target zones at the pial surface earlier in development?

The directional growth of a columnar/radial glial cell resembles the axonal guidance of a young CNS neuron towards its target cell. The developing columnar/radial glial cells and young neurons in the CNS have also anatomical similarities. (a) Their cell bodies produce elongating processes, i.e. radial fibers and axons, respectively. (b) These processes have cytoskeletons of axial microtubules and filaments. (c) The terminal end-knobs in peripheral branches of the radial fibers resemble the growth cones of young neurites (Cajal 1909), as first pointed out by Hinds and Ruffett (1971). These similarities imply that conclusions from studies of growing neurites, especially in tissue culture, can shed light on the directional guidance of radial fibers in the intact brain as well. Such studies have shown that microtubules are necessary for the growth and maintenance of axonal shape and that the growth cones are the leading edges in the growing processes and respond to environmental stimuli (e.g. NGF) and guide the neurites towards their targets (Landis 1983; Bunge 1986; Lockerbie 1987). Future *in vitro* studies of factors influencing growth and guidance of columnar and radial glial cells may likewise better characterize these poorly understood processes. Of particular interest would be studies of the end knobs which, like growth cones, can be isolated in a test tube and identified in the electron microscope (Pfenninger et al. 1983). When the end feet and end knobs of the columnar cells reach the basal lamina they become flattened against it, suggesting adhesion (Figs. 13, 14). The basal aspects of cells with short outer processes likewise seem to adhere to the inside of the basal lamina (Figs. 5, 6). The term "end feet" is appropriate since the basal lamina provides a foothold for them (Hay 1973; Hogan 1981; Vracho 1982).

It was noted in cultures of sympathetic neurons that the points of attachment are between their growth cones and the substrate (Bray 1973). Similarly, the end feet of the epithelial cells adhere selectively to their substrate, the basal lamina. Thus, a columnar/radial glial cell has one anchor point at the apical junctions and another at the basal lamina, which has been shown to function as a scaffolding for epithelial cells during development. The basal lamina may also be involved in the migration and differentiation of such cells (Sanders 1983).

The columnar cell collapse, division, and elongation that are associated with nuclear movement (Fig. 29), imply adhesion changes at the basal lamina where cells repeatedly are detached and reattached. Since end knobs form a continuous external limiting membrane (Fig. 13), the end knobs of cells, which are detached while preparing for mitosis must be replaced by other processes' end knobs, which are moving towards the basal lamina after mitosis. Other sources of replacements are sprouts from cells which already are attached to the basal lamina. It is likely that the terminal branches of the columnar cells constitute a flexible system, i.e. they develop, divide, sprout, and attach themselves in ways which are regulated locally by the extracellular environment.

Membrane changes may be involved in locomotion, growth and rebuilding of the end feet. The absence of microtubules indicates that the end feet can adapt to local stimuli rapidly by changing shape and adhesive properties. Autophagic segregation with degradation of cytoplasm (Fig. 13B), a mechanism for remodelling of the end-feet, has also been seen in the growth cones of axons (Bunge 1973) and in the end feet of radial fibers (Rakic 1972). The end knobs contain mitochondria, vesicles, and profiles of SER (Fig. 13). It is probable that the small vesicles emanating from the SER (Fig. 12B) or the surface membrane merge to form larger vesicles (Fig. 12E), or are inserted into the plasmalemma to increase its surface area (Fig. 12C, D). There is evidence to support this concept of focal plasmalemmal growth in neuronal growth cones (Pfenninger and Bunge 1974). It cannot be excluded, however, that such subplasmalemmal vesicles result from the preparative procedure (Landis 1983).

6.4 Organogenesis of the Telencephalic Wall

Morphogenesis of the neopallial wall, which in the preneuronal period consists of a monolayer of epithelial cells (Fig. 2), involves changes in three parameters, the curvature, the thickness, and the surface area of this cohesive sheet of cells.

The telencephalic hemispheres are formed by *invagination* of the neuroepithelium in the forebrain (Fig. 2). Invagination of epithelial sheets, an important feature of morphogenesis in all organs (Balinsky 1981), has been studied extensively, mostly in nonmammalian species (Ettensohn 1985). It is the result of a mechanical process that includes several well coordinated forces. Neurulation and the subsequent maturation of the CNS are major examples of developmental invagination in the CNS (Jacobson 1981; Gordon 1985).

Studies in developing chicken embryos have shown that CSF pressure inflates the hemispheres like balloons, and that a reduction of pressure is associated not only with collapse of the hemispheres but also with retarded growth of their nerve cells (Jelinek and Pexieder 1970; Pexieder and Jelinek 1970; Desmond and Jacobson 1977; Pacheco et al. 1986). The conclusion is that CSF pressure is needed for normal morphogenesis of the telencephalic hemispheres and proper growth and development of the neuroepithelial cells. In mouse embryos closure of the brain is associated with elongation of the prosencephalon (Jacobson and Tam 1982). During the course of the present investigation of fetal rat it was found that a puncture of an unfixed hemisphere (after sacrifice) at E11–13 always resulted in its collapse (Åström and Webster 1990). These observations support the notion that CSF pressure also inflates the telencephalic vesicles in rodents during a certain stage of development, but further studies of this phenomenon are needed in these animals.

The following anatomical and physiological features appear to be essential for whatever mechanism causes the inflation of the hemispheres and the radial alignment of epithelial cells.

1. The inner ends of the bipolar and mitotic cells are joined by junctional complexes and form the ventricular surface, which has structural integrity, has the appearance of a mosaic in tangential sections (Sauer 1937), and is probably also a diffusion barrier.
2. The fluid compartment of the brain becomes closed during development; in chicken embryos closure of the CSF compartment of the brain and rostral spinal cord coincides with the onset of rapid brain enlargement (Desmond and Jacobson 1977).
3. Fluid enters the CSF space. The production of ventricular fluid will be discussed in Chap. 7.
4. The cell monolayer forming the telencephalic wall is thin and therefore flexible.

Hydrostatic pressure applied to a homogeneous monolayer of cells would only produce one single cavity. In order to explain the actual configuration of the telencephalon, which has two hemispheres separated by an interhemispheric fissure (Fig. 2), it is necessary to postulate that the effect of CSF pressure is modified by gradients in cell adhesion. (Gustafsson and Wolpert 1967; Nardi 1981; Jacobson 1985). Tangential pressure produced by cell mitosis (Ettensohn 1985) may also modify the effect of radially directed pressure produced by the CSF. During this period of development (E11–13) the number of cells is increasing rapidly in the telencephalic wall but is more or less constant in the telencephalic roof where the rate of cell division is slower and where frequent spontaneous cell necrosis occurs.

A change in the shape of cells is frequently cited as a major factor in invagination. Growth of axial microtubules and contraction of microfilaments produces elongation and reduction of the transverse diameter of columnar cells (Burnside 1971, 1973). The contraction of microfilaments with narrowing of the apical ends of cells has been correlated especially with the bending of an epithelial sheet, as seen during neurulation (Baker and Schroeder 1967; Wessells et al. 1971; Sadler et al. 1982). Coordinated contractions of the apical ends will bend the neuroepithelial plate outwards. Since the telencephalic wall has the same basic structure as the neural plate and wall, it is possible that the contraction of microfilaments at the ventricular end of these cells likewise contributes to the indentation which produces the vesicles.

In conclusion, the telencephalic hemispheres are initially shaped by a number of forces which are precisely coordinated in time and place. CSF pressure of the required magnitude (Pexieder and Jelinek 1970) is produced by filtration and/or secretion of fluid into a ventricular space which is closed at the appropriate stage of development. Other factors of possible significance are gradients in adhesion between cells, regional differences in the proliferation of cells, the shapes of cells, and the elasticity of a wall which varies in thickness. The rapid shape changes that occur in the telencephalic vesicles are of crucial importance for the development of the cerebrum. They deserve further study in mammalian species.

The *thickness of the telencephalic wall* is related to the height of individual columnar cells in the preneural period and to the height of individual radial glial cells during the period of early neuronal migration. It has been shown in urodeles that height and height increases are intrinsic properties of the cells in the neural tube (Holtfreter 1947; Burnside 1971). Assuming that the cells in the early telencephalic wall of rodents have similar properties, one can postulate that each epithelial cell has a "height program" (Jacobson and Gordon 1976). These programs must be coordinated spatially and temporally, since at any given stage of development the telencephalic wall tapers in a lateromedial direction (Fig. 2), and since each new generation of cells is taller than the preceding one. In other words, a columnar cell in the telencephalic wall must "know" its height in relation to that of its neighbors, and to its location in the hemisphere, and the "knowledge" regulating the next increment of growth must be transferrable to the next generation.

The *surface area of the telencephalic wall* increases during preneuronal development due to the proliferation of cells. The new cells step into the line of columnar cells, all of which are attached to each other at the ventricular surface. No cells separate and move away from the ventricular surface until neurons appear.

6.5 Functions of Columnar Cells

6.5.1 Mechanical Functions

A major function of columnar cells is to give shape to CNS in the preneuronal period.

The apical ends of the columnar cells constitute the ventricular surface. Their closely packed inner processes, united by impermeable junctional complexes, probably limit the escape of fluid from the ventricles and create the increase of intraventricular pressure which is needed for inflation of the hemispheres and growth of the brain.

The end feet and end knobs of the bipolar cells form together an "external limiting membrane" (Hinds and Ruffett 1971), which is covered by a basal lamina. These structures constitute the pia-glia surface of the early CNS. As the outer processes of the columnar cells grow in length during E11–12, the marginal zone, which is free of nuclei, will increase in width. Late in E12 or early in E13 the first nerve cells (Fig. 14) will settle in this superficial zone, where layer I in the 6-layered cortex later will appear (Åström 1967). These cells make up the first stratum of cells within the neopallial wall, which up to this stage of development has only been pseudostratified (Smart and McSherry 1982). The external limiting membrane prevents the nerve cells in layer I from directly contacting the pial surface. In later development glial cells likewise will isolate neurons from direct contact with each other (except at synapses), with blood vessels, and with surfaces of the CNS.

During the period of neuronal migration the columnar cells become replaced by radial glial cells. These cells form the scaffolding in which neurons and astroglial cells of more mature type are embedded. In addition, their processes guide migrating nerve cells (Rakic 1971, 1972) and provide a framework where young neurons can settle before developing axons and dendrites (Hatten and Liem 1981).

In conclusion, the columnar/radial glial cells determine the form of the CNS upon which all further morphogenesis is patterned and provide a structural framework which is essential for correct positioning of neurons and their growing processes. As development progresses, these cells are replaced by phenotypically more mature glial cells, many more neurons appear and, after the neurons migrate to predetermined locations, the brain is gradually transformed into an organ of increasing size and complexity. Nevertheless, the shape of the growing brain can be traced back to and is defined by the arrangement of columnar cells during the preneuronal period. The framework formed by these cells and the radial glial cells is temporary and it is dismantled when it is no longer needed, i.e. after the CNS has acquired its basic shape, after all nerve cells have been guided to their destinations, and after the other functions of the radial glial cells have been taken over by more specialized glial cells. Ependymal cells will separate the CNS parenchyma from the fluid filled ventricles and astrocytes will form the glial membrane at the pial surface and give support to the CNS through their processes and perivascular attachments.

6.5.2 *Metabolic Functions*

The structure of the columnar cell in the telencephalic wall reflects a comparatively low level of metabolism compared to that of cells in the area choroidea. The most important function is probably synthesis of macromolecules for development and growth of the radial processes. It is mainly carried out in the perinuclear region, which contains almost all ribosomes, and in the proximal part of the inner process, where the Golgi apparatus is located. Axially oriented profiles of SER probably have a transport function.

Blood vessels do not invade the telencephalic wall until late E12. Up to that time, the exchange of nutrients and metabolites must take place through the basal lamina and between the end-knobs of the bipolar cells.

6.5.3 *Germinal Functions*

It is remarkable that, during interphase, germinal cells become differentiated into columnar cells with specialized functions and that these cells then can return to the undifferentiated state of mitotic cells. It is evident that this arrangement is "economical" for the organism, at least up to a certain stage of development.

7 Discussion: Area Choroidea

The cells in the telencephalic roof, like those in the neopallial wall, give shape to the early brain, provide a framework for future development, limit the escape of CSF, and harbor a few germinal cells. We will now turn to features which are characteristic for cells in the two locations.

7.1 Epithelial Polarity

Since the telencephalic wall is a cohesive monolayer of epithelial cells, it probably shares some of the functions of such structures in other organs. Epithelium in general is a barrier between internal and external regions of the body. It is involved in the structure of organs, movement of fluids, secretion, absorption, and regulation of homeostasis (Berridge and Oschmann 1972). The basolateral parts of cells facing mesenchymal tissue with blood vessels and fluid-filled intercellular compartments are involved in metabolic functions and have receptors for hormonal signals. The apical ends facing the "external milieu," which frequently is a fluid-filled compartment, secrete products and fluid; apical ends also absorb material from bodily cavities. These vectorial functions are reflected in the polarity of epithelial cells, a topic which has received much attention in recent years (Rodriguez-Boulan 1983; Simons and Fuller 1985; Handler 1989; Nelson 1989; Rodriguez-Boulan and Nelson 1989). Research on the function of polarized cells has mainly been carried out in cultured cells, especially those of kidney tubules (Matlin and Valentich 1989). Of particular interest is the fact that the membranes over the baso-lateral and apical poles are specialized to carry out different vectorial functions, and that the distribution of cytoskeletal elements and organelles also is polarized and is important in their structure and function. Similar investigations of functions of the telencephalic wall have not been carried out. Comparisons between the morphology of this and other epithelial organs and data from cultured cells, however, suggest interpretations which may provide a basis for direct studies of the function of neuroepithelial cells in general and those in the early telencephalic wall in particular.

7.2 Metabolic Functions

We have earlier stated that the columnar/radial glial cells in the telencephalon mainly have structural functions, i.e. they give shape to the growing brain and guide migrating neurons. We concluded that the internal structure of these cells reflects a metabolism which is low; it is necessary mainly for survival, division, development and growth of these cells.

In contrast, the cells in the area choroidea have a cuboidal or low columnar shape which does not change during development and their mitotic index appears to be low. They are similar in appearance to metabolically active epithelial cells elsewhere in the body, e.g. in the kidney tubules, intestinal mucosa, glands, and choroid plexus. We suggest that the higher metabolic level of the roof cells is needed more for specialized epithelial functions such as transport, absorption and secretion, than for growth and structural maintenance. These functions are reflected in the electron microscopic appearance of the cells (Fig. 28). Interdigitations of the basal processes beneath the basal lamina increase the area for absorption of fluid and exchange of metabolities. The large Golgi apparatus and the abundance of organelles in the inner portion of a cell are commensurate with high metabolic activity. The fat droplets characteristically found in these cells may provide energy for this activity. Finally, the apical protrusions increase the surface area for absorption and discharge of material and fluid from and to the CSF.

7.3 Transport of Fluid

A major function of the area choroidea is probably the transport of fluid across its epithelium. The area choroidea, after E13, gives rise to the choroid plexus in the lateral and third ventricles. Furthermore, up to E13, the cells in the area choroidea are very similar to the epithelial cells which will appear later in the choroid plexus during an early phase of their development (Tennyson and Pappas 1964, 1967). The main similarities are as follows: convex side toward the ventricle (less in choroid epithelium), irregular villi on the surface (less in roof cells), micropinocytosis at the surface, more organelles in inner than in outer processes of cells, infolding and interdigitation of basal processes. There are also differences: the roof cells contain fat droplets which are never seen in the choroid plexus at any stage of development, glycogen particles appear in the epithelium of the choroid plexus at a later stage of development, but are never seen in the roof cells. These data suggest that the area choroidea produces CSF fluid before the true choroid plexus has developed.

7.4 Area Choroidea as a Gland

The question could be raised whether the area choroidea has the function of a gland during fetal life for the following reasons: (a) It is located where the paraphysis appears in lower animals and where a rudiment of this gland is seen in some mammals during fetal life (Warren 1917; Ariens Kappers 1955). (b) True glands, notably the subfornical organ (Akert et al. 1961) and the subcommissural organ (Rakic and Sidman 1968), are located in the dorsal midline of the fetal brain on either side of area choroidea. (c) The general structure of the roof cells and especially the abundance of organelles in their apical parts are compatible

64

with gland-like functions. It is true that a considerable amount of material from necrotic cells (Graumann 1950) is discharged from the area choroidea into the ventricular system at this stage of development, and that biologically active molecules are gradually released from these cells and fragments. Thus, this region has a gland-like function in the sense that biological material is "secreted" from it. However, histological signs of true secretion e.g. secretory vesicles (Fawcett 1986) are not seen here.

7.5 Absorptive Functions

The apical ends of the roof cells protrude into the ventricle in bulblike formations resulting in an enlargement of the surface area that is exposed to the CSF. This suggests that the area choroidea has an absorptive function. The existence of numerous intermediate phases of micropinocytosis at the surface membrane (Fig. 25) further supports the view that absorption is taking place. These intermediate steps are: focal alteration of the surface with thickening of plasmalemma, adhesion of material and bristle-like projections towards the cytoplasm; invagination of such areas; the formation of coated vesicles (Fig. 24C). The fact that vesicles do not accumulate at the surface and that the number of vesicles there usually does not exceed that of the patches and invaginations suggests that this movement is a smooth continuous process. Micropinocytosis with coated vesicles is a well described biological phenomenon. It has been noted in erythroblasts of normal bone marrow (Fawcett 1965), sinusoidal surfaces of liver cells (Roth and Porter 1962), Purkinje cells of the cerebellum (Palay 1963), nerve cells and neuroglia (Andres 1964; Holtzman and Peterson 1969), renal epithelium (Ericsson 1964), and the vas deferens (Friend and Farquhar 1967). Its purpose is to bring material, especially protein, into cells. In the case of the roof cells, this material is usually seen as small particles which probably have emanated from disintegrating macrophages in the same region.

The following observations in this study suggest a relationship between coated vesicles and MVBs:

1. Mergers between coated vesicles and MVBs have been observed (Figs. 25B,C). A similar observation was made by Friend and Farquhar (1967).
2. The walls of the coated vesicles resemble the plaque-like densities in the MVBs (Figs. 21 and 25A, D). These plaques were also noted by Friend and Farquhar.
3. Coated vesicles and MVBs are most commonly seen in the same region of the cell, i.e. the apical portion near the ventricular surface.

These observations have led to the conclusion that some, if not all, coated vesicles merge with and become part of MVBs. This raises the question whether a MVB receives its membranous material in part from the apical surface and in part from other structures, e.g. the Golgi-complex, or whether its membranes are formed exclusively at the apical surface. It the latter possibility is correct, one is left with the question, how is the membrane which is lost through micropinocytosis replaced? Although this study does not provide quantitative data, a general impression is that the micropinocytotic processes lead to considerable loss of surface membrane in the bulbous protrusions.

The spatial relationship that exists between small Golgi vesicles and MVBs (Fig. 25B) suggests that the former are primary lysosomes which carry enzymes to the latter, which then function as secondary lysosomes.

The special shape of roof cells and the large number of micropinocytotic vesicles at their surfaces suggest that the absorption occurring in this region is an important function of the roof cell. The purpose is not clear. The uptake of material may represent the ultimate disposal of necrotic cells and other fragments in the ventricle, or it may indicate selective absorption of material that serves special needs of the cell, or both.

8 Summary

The telencephalic wall was studied by light and electron microscopy in 11–13 day old fetal rats (E11–13). A few specimens from E14–16 were also included for comparisons. Two areas were selected: the dorso-lateral convexity of the hemispheric vesicles, called the neopallial wall, and the area choroidea, the posterior part of the telencephalic roof which unites the two hemispheres.

Our observations and a review of the literature have shown that on E11–12 the neopallial wall, the telencephalic roof, and the hippocampal anlage between them form a continuous, nonstratified, cohesive monolayer of columnar and mitotic cells, which essentially is similar to epithelial monolayers elsewhere in the body. This simple structure is modified late on E12 or early in E13 in the neopallial wall when postmitotic neurons appear and migrate in order to form the cortical plate. However, bipolar radially oriented cells, which span the entire width of the wall, still predominate. These cells, now called *radial glial cells*, increase greatly in number and length during the period of neuronal migration.

The *cuboidal* cells in the neural tube, the *columnar* cells in the early neopallial wall, and the *radial glial cells* in the period of neuronal migration have the same basic structure. They are axially polarized epithelial cells which are characterized by the following basic features. They have an elongated bipolar shape which is maintained by a cytoskeleton of longitudinally oriented microtubules. Opposite ends are different structurally and functionally. Thus, the apical ends, connected by tight junctions, face the fluid-filled cavity while the outer ends, covered by a basal lamina, face mesenchymal tissue including blood vessels. A polarization of cytoplasmic organelles is also evident, e.g. the Golgi apparatus has always a supranuclear position. During the early development of the telencephalon this basic epithelial structure is maintained but is modified locally in order to serve various functions. The columnar/radial glial cells in the neopallial wall are elongated and slender, have a narrow Golgi apparatus, profiles of RER and vesicles, relatively few ribosomes, and show a few examples of micropinocytosis. These cells grow continuously in length during development. On the other hand, the cells in the area choroidea have a low columnar or cuboidal shape, which does not change during development. The inner portion (between the nucleus and the ventricle) contains a voluminous Golgi apparatus, many mitochondria, RER cisternae which contain electron-dense material, SER, and many vesicles. The inner ends of the cells project into the ventriclar cavity as bulbous or apical protrusions which contain many organelles, especially MVBs. Micropinocytosis is common on or near the surface membrane. A unique feature of cells in the area choroidea is the presence of cytoplasmic bodies which have the appearance of lipid droplets; they are seen in all parts of the cells.

On the basis of the microscopic appearance as described in this paper and the experimental work carried out by others, we suggest that the main function of the columnar cells in the early neopallial wall is structural. These cells seal off the ventricular cavity which contains CSF, they form an external limiting membrane at the pial surface, they give shape to the growing telencephalon thereby determining the future form of the brain, and they set the stage for positioning of the neurons when they appear. The columnar cells form a temporary structure, a scaffolding, which is dismantled when no longer needed. These cells are then replaced by more specialized neuroglial cells.

The structure of the columnar cells in the neopallial wall reflects a comparatively low level of metabolism in contrast to those in the area choroidea which are similar to metabolically active epithelial cells elsewhere, e.g. in the kidney tubules. We suggest that the metabolism of the cells in area choroidea is important particularly for specialized epithelial functions, e.g. transport of fluid and substances and absorption from the CSF.

Other topics that have been discussed are: mechanisms for change of shape of dividing cells during interphase, directional movements of postmitotic columnar and radial glial cells, and organogenesis of the telencephalic hemispheres. We hope that our observations and discussion are useful for those studying CNS *in vivo* development by using cultured cells and explants, by using methods to define cell lineages, and by direct microscopic observation of living animals in which dynamic events in the CNS can be studied directly.

Acknowledgements

This study was supported in part by grant N.S.07596 from the National Institute of Neurological Disorders and Stroke. Some of the work was carried out in the Departments of Neurology and Neuropathology at Massachusetts General Hospital, Boston, Massachusetts, and the Eunice Kennedy Shriver Institute, Waltham, Massachusetts. We gratefully acknowledge the photographic help of Mr. L. Cherkas, the secretarial help of Ms. S. Wentzel, and the technical help of Ms. Y. Chang. We also want to thank Dr. Raymond D. Adams for his support and encouragement.

References

Akert K, Potter HD, Anderson JW (1961) The subfornical organ in mammals. I. Comparative and topographical anatomy. J Comp Neurol 113: 1–13

Andres KH (1964) Micropinocytosis im Zentralnervensystem. Z Zellforsch 64: 63–73

Ariens Kappers J (1955) Development of human paraphysis. J Comp Neurol 102: 425–498

Åström KE (1967) On the early development of the isocortex in fetal sheep. Prog Brain Res 26: 1–59

Åström KE, Webster H deF (1990) Preparation of fetal rat brains for light and electron microscopy. J Electron Microsc Tech 15: 383–396

Bailey P (1916) Morphology of the roof-plate of the forebrain and the lateral choroid plexus in the human embryo. J Comp Neurol 26: 79–120

Baker P, Schroeder TE (1967) Cytoplasmic filaments and morphogenetic movement in the amphibian neural tube. Dev Biol 15: 432–450

Balinsky BI (1981) An introduction to embryology 5th edn, Saunders, Philadelphia

Bayers B, Porter KR (1964) Oriented microtubules in elongating cells of the developing lens rudiment after induction. Proc Natl Acad Sci USA 52: 1091–1099

Berridge MJ, Oschmann JL (1972) Transporting epithelia. Academic, New York

Bignami A, Dahl D (1974) Astrocyte-specific protein and neuroglial differentiation. An immunofluorescence study with antibodies to the glial fibrillary acidic protein. J Comp Neurol 153: 27–38

Boulder Committee (1970) Embryonic vertebrate central nervous system: revised terminology. Anat Rec 166: 257–262

Bray D (1973) Branching patterns of individual sympathetic neurons in culture. J Cell Biol 56: 702–712

Bunge MB (1973) Fine structure of nerve fibers and growth cones of isolated sympathetic neurons in culture. J Cell Biol 56: 713–735

Bunge MB (1986) The axonal cytoskeleton: its role in generating and maintaining cell form. Trends Neurosci 9: 477–482

Burnside B (1971) Microtubules and microfilaments in newt neurulation. Dev Biol 26: 416–441

Burnside B (1973) Microtubules and microfilaments in amphibian neurulation. Am Zool 13: 989–1006

Burnside B (1975) The form and arrangement of microtubules: an historical, primarily morphological review. Am N Y Acad Sci 253: 14–26

Cajal S Ramon Y (1909) Histologie du système nerveux de l'homme et des vertébrés. Vol 1 ch 21. Maloine, Paris

Choi BH (1988) Prenatal gliogenesis in the developing cerebrum of the mouse. Glia 1: 308–316

Choi BH, Lapham LW (1978) Radial glia in the human fetal cerebrum: a combined Golgi, immunofluorescent and electron microscopic study. Brain Res 148: 295–311

Dahl D (1981) The vimentin-GFA protein in rat neuroglia cytoskeleton occurs at the time of myelination. J Neurosci Res 6: 741–748

Dahl D, Rueger DC, Bignami A (1981) Vimentin, the 57000 molecular weight protein of fibroblast filaments, is the major cytoskeletal component in immature glia. Eur J Cell Biol 24: 191–196

Desmond ME, Jacobson AG (1977) Embryonic brain enlargement requires cerebrospinal fluid pressure. Dev Biol 57: 188–198

Duncan D (1957) Electron microscope study of the embryonic neural tube and notochord. Tex Rep Biol Med 15: 367–377

Eckenhoff MF, Rakic P (1984) Radial organization of the hippocampal dentate gyrus: a Golgi, ultrastructural, and immunocytochemical analysis in the developing rhesus monkey. J Comp Neurol 223: 1–21

Ericsson JL (1964) Absorption and decomposition of homologous hemoglobin in renal proximal tubular cells; an experimental light and electron microscopic study. Acta Pathol Microbiol Scand Suppl 168: 1–121

Ericsson JL (1969) Mechanism of cellular autophagy. In: Dingle JT, Fell HB (eds) Lysosomes in biology and pathology, vol 2. North Holland Publishing, Amsterdam, pp 345–394

Ettensohn CA (1985) Mechanisms of epithelial invagination. Q Rev Biol 60: 289–307

Fawcett DW (1965) Surface specialization of absorbing cells. J Histochem Cytochem 13: 75–91

Fawcett DW (1981) The cell, 2nd ed. Saunders, Philadelphia

Fawcett DW (1986) Bloom and Fawcett, A textbook of histology, 11th ed. Saunders, Philadelphia

Fedoroff S (1986) Prenatal ontogenesis of astrocytes. In: Fedoroff S, Vernadakis A (eds) Astrocytes. Development, morphology, and regional specialization of astrocytes. Academic, New York, pp 35–74 (Cellular neurobiology, vol 1)

Friend DS, Farquhar MG (1967) Functions of coated vesicles during protein absorption in the rat vas deferens. J Cell Biol 35: 357–376

Fujita S (1963) The matrix cell and cytogenesis in the developing central nervous system. J Comp Neurol 129: 37–42

Gadisseux J-F, Evrard P (1985) Glial-neuronal relationship in the developing central nervous system. A histochemical-electron microscope study of radial glial cell particulate glycogen in normal and reeler mice and the human fetus. Dev Neurosci 7: 12–32

Gordon R (1985) A review of the theories of vertebrate neurulation and their relationship to the mechanics of neural tube birth defects. J Embryol Exp Morph Suppl 89: 229–255

Graumann W (1950) Zelldegeneration im Telencephalon medium und Paraphysen-Entwicklung bei der weissen Maus. Z Anat Entwickl-Gesch 115: 19–31

Gustafsson T, Wolpert L (1967) Cellular movement and contact in sea urchin morphogenesis. Biol Rev 42: 442–498

Handler JS (1989) Overview of epithelial polarity. In: Eaton DC (ed) Special topic: polarity of epithelial cells: intracellular sorting and insertion. Ann Rev Physiol 51: 729–740

Hatten ME, Liem RKH (1981) Astroglial cells provide a template for the positioning of developing cerebellar neurons in vitro. J Cell Biol 90: 622–630

Hay E (1973) Role of basement membranes in development and differentiation. In: Kefalides NA (ed) Biology and chemistry of basement membranes. Academic, New York, pp 119–136

Herman L, Kauffman SL (1966) The fine structure of the embryonic mouse neural tube with special reference to cytoplasmic microtubules. Dev Biol 13: 145–162

Hinds JW, Ruffett TL (1971) Cell proliferation in the neural tube: an electron microscopic and Golgi analysis in the mouse cerebral vesicle. Z Zellforsch 115: 226–264

Hines M (1922) Studies on the growth and differentiation of the telencephalon in man. The fissura hippocampi. J Comp Neurol 34: 73–171

His W (1889) Die Neuroblasten und deren Entstehung im embryonalen Mark. Abhandl Konigl Sachs Ges Wissensch. Math Phys Kl. 15: 313–372

Hockfield S, McKay RDG (1985) Identification of major cell classes in the developing mammalian nervous system. J Neurosci 5: 3310–3328

Hogan B (1981) Laminin and epithelial cell attachment. Nature 290: 737–738

Holtfreter J (1947) Structure, motility and locomotion in isolated embryonic amphibian cells. J Morphol 79: 27–62

Holtzman E, Peterson ER (1969) Uptake of protein by mammalian neurons. J Cell Biol 40: 863–869

Houle J, Fedoroff S (1983) Temporal relationship between the appearance of vimentin and neural tube development. Dev Brain Res 9: 189–195

Jacobson AG (1981) Morphogenesis of the neural plate and tube. In: Connelly TG, Brinkley LL, Carlson BM (eds) Morphogenesis and pattern formation. Raven, New York, pp 233–264

Jacobson AG (1985) Adhesion and movement of cells may be coupled to produce neurulation. In: Edelman GM, Thiery JP (eds) The cell in contact. Adhesions and junctions as morphogenetic determinants. Wiley, New York, pp 49–65

Jacobson AG, Gordon R (1976) Changes in the shape of the developing vertebrate nervous system analyzed experimentally, mathematically and by computer simulation. J Exp Zool 197: 191–246

Jacobson AG, Tam PPL (1982) Cephalic neurulation in the mouse embryo analyzed by SEM and morphometry. Anat Rec 203: 375–396

Jelinek R, Pexieder T (1970) Pressure of the CSF and the morphogenesis of the CNS. I. Chick. Folia Morphol 18: 102–110

Karfunkel P (1971) The role of microtubules and microfilaments in neurulation in *Xenopus*. Dev Biol 25: 30–56

Landis S (1983) Neuronal growth cones. Ann Rev Physiol 45: 567–580

Levitt P, Rakic P (1980) Immunoperoxidase localization of glial fibrillary acidic protein in radial glial cells and astrocytes of the developing rhesus monkey brain. J Comp Neurol 193: 815–840

Lockerbie RO (1987) The neural growth cone: a review of its locomotory, navigational, and target recognition capabilities. Neuroscience 20: 719–729

Lorente de Nó R (1933) Architectonics and structure of the cerebral cortex. J Psychol Neurol (Lpz) 45: 381–438

Lyser KM (1964) Early differentiation of motor neuroblasts in the chick embryo as studied by electron microscopy. I. General aspects. Dev Biol 10: 433–466

Lyser KM (1968) Early differentiation of motor neuroblasts in the chick embryo as studied by electron microscopy. II. Microtubules and neurofilaments. Dev Biol 17: 117–142

Matlin KS, Valentich JD (eds) (1989) Functional epithelial cells in culture. Liss, New York

Matsuyama H, Komatsu N, Senda R (1973) Electron microscopic studies on the developing telencephalic wall of the rat fetus. Okajimas Folia Anat Jpn 50: 273–294

Meller K, Breipohl W, Glees P (1966) Early cytological differentiation in the cerebral hemisphere of mice. Z Zellforsch 72: 525–533

Mountcastle VB (1979) An organizing principle for cerebral function: the unit module and the distributed system. In: Schmitt FO, Worden FG (eds) The neurosciences. Fourth study program. MIT Press, Cambridge, pp 21–42

Nardi JB (1981) Epithelial invagination: adhesive properties of cells can govern position and directionality of epithelial folding. Differentiation 20: 97–103

Nelson WJ (1989) Development and maintenance of epithelial polarity: a role for the submembranous cytoskeleton. In: Matlin KS, Valentich JD (eds) Functional epithelial cells in culture. Liss, New York, pp 3–42 (Modern cell biology, vol 8)

Nowakowski RS (1987) Basic concepts of CNS development. Child Dev 58: 568–595

Pacheco MA, Marks RW, Schoenwolf GC, Desmond ME (1986) Quantification of the initial phases of rapid brain enlargement in the chick embryo. Am J Anat 175: 403–411

Palay SL (1963) Alveolate vesicles in Purkinje cells of the rat's cerebellum. J Cell Biol 19: 89A

Peters A, Feldman M (1973) The cortical plate and molecular layer of the late rat fetus. Z Anat Entwickl-Gesch 141: 3–37

Pexieder T, Jelinek R (1970) Pressure of the CNS and the morphogenesis of the CNS. II. Pressure necessary for normal development of brain vesicles. Folia Morphol 18: 181–192

Pfenninger KG, Bunge RP (1974) Freeze-fracturing of nerve growth cones and young fibers. A study of developing plasma membrane. J Cell Biol 63: 180–196

Pfenninger KH, Ellis L, Johnson MP, Friedman LB, Somlo S (1983) Nerve growth cones isolated from fetal rat brain: subcellular fractionation and characterization. Cell 35: 573–584

Pixley SKR, De Vellis J (1984) Transition between immature radial glia and mature astrocytes studied with a monoclonal antibody to vimentin. Dev Brain Res 15: 201–209

Porter KR (1973) Microtubules in intracellular locomotion. In: Abercrombie M (ed) Locomotion of tissue cells. Elsevier, Amsterdam, pp 149–169 (Ciba Foundation Symposium)

Rakic P (1971) Guidance of neurons migrating to the fetal monkey neocortex. Brain Res 33: 471–476

Rakic P (1972) Mode of cell migration to the surperficial layers of fetal monkey neocortex. J Comp Neurol 145: 61–84

Rakic P (1978) Neuronal migration and contact guidance in the primate telencephalon. Postgrad Med J 54: 25–40

Rakic P (1982) Early developmental events: cell lineages, acquisition of neuronal positions, and area and laminar development. Neurosci Res Prog Bull 20: 439–451

Rakic P, Sidman RL (1968) Subcommissural organ and adjacent ependyma: autoradiographic study of their origin in the mouse brain. Am J Anat 122: 317–335

Retzius G (1893) Studien Über Ependym und Neuroglia. Biol Untersuch (Stockholm) 5: 9–26

Retzius G (1894) Die Neuroglia des Gehirns beim Menschen und bei Säugethieren. Biol Untersuch (Stockholm) 6: 1–24

Rickmann M, Wolff JR (1985) Prenatal gliogenesis in the neopallium of the rat. Adv Anat Embryol Cell Biol 93: 1–104

Rodriguez-Boulan E (1983) Membrane biogenesis, enveloped RNA, viruses, and epithelial polarity. In: Satir BH (ed) Modern cell biology, vol 1. Liss, New York, pp 119–170

Rodriguez-Boulan E, Nelson WJ (1989) Morphogenesis of the polarized epithelial cell phenotype. Science 245: 718–725

Roth TF, Porter KR (1962) Specialized sites on the cell surface for protein uptake. In: Breese SS (ed) Electron microscopy. 5th International Congress. Academic, New York

Sadler TW, Greenberg P, Coughlin P, Lessard JL (1982) Actin distribution patterns in the mouse neural tube during neurulation. Science 215: 172–174

Sanders EJ (1983) Recent progress towards understanding the roles of the basement membranes in development. Can J Biochem Cell Biol 61: 949–956

Sauer FC (1935a) Mitosis in the neural tube. J Comp Neurol 62: 377–405

Sauer FC (1935b) The cellular structure of the neural tube. J Comp Neurol 63: 13–23

Sauer FC (1936) The interkinetic migration of embryonic epithelial nuclei. J Morphol 60: 1–11

Sauer FC (1937) Some factors in the morphogenesis of vertebrate embryonic epithelia. J Morphol 61: 563–579

Sauer ME, Chittenden AC (1959) Deoxyribonucleic acid content of cell nuclei in the neural tube of the chick embryo: evidence for intermitotic migration of nuclei. Exp Cell Res 16: 1–16

Sauer ME, Walker BE (1959) Radioautographic study of interkinetic nuclear migration in the neural tube. Proc Soc Exp Biol Med 101: 557–560

Schmechel DE, Rakic P (1979) A Golgi study of radial glial cells in developing monkey telencephalon: morphogenesis and transformation into astrocytes. Anat Embryol 156: 115–152

Schnitzer J, Franke WW, Schachner M (1981) Immunocytochemical demonstration of vimentin in astrocytes and ependymal cells of developing and adult mouse nervous system. J Cell Biol 90: 435–447

Seymour RM, Berry M (1975) Scanning and transmission electron microscope studies of interkinetic nuclear migration in the cerebral vesicles of the rat. J Comp Neurol 160: 105–126

Sidman RL, Miale IL, Feder N (1959) Cell proliferation and migration in the primitive ependymal zone; an autoradiographic study of histogenesis in the nervous system. Exp Neurol 1: 322–333

Simons K, Fuller S (1989) Cell surface polarity in epithelia. Ann Rev Cell Biol 1: 243–288

Smart IHM, McSherry GM (1982) Growth patterns in the lateral wall of the mouse telencephalon. II. Histological changes during and subsequent to the period of isocortical neuron production. J Anat 134: 415–442

Stensaas LJ, Stensaas SS (1968) An electron microscopic study of cells in the matrix and intermediate laminae of the cerebral hemisphere of the 45 mm rabbit embryo. Z Zellforsch 91: 341–325

Tennyson VM, Pappas GD (1964) Fine structure of the developing telencephalic and myelencephalic choroid plexus in the rabbit. J Comp Neurol 123: 379–412

Tennyson VM, Pappas GD (1967) The fine structure of the choroid plexus: adult and developmental stages. In: Lajtha A, Ford DH (eds) Brain barrier systems. Prog Brain Res 29: 63–85

Trinkaus JP (1982) Some thoughts on directional cell movement during morphogenesis. In: Bellairs R, Curtis A, Dunn G (eds) Cell behaviour. Cambridge University Press London, pp 471–498

Trinkaus JP (1984) Cells into organs. The forces that shape the embryo, 2nd edn. Prentice Hall, New York

Valentino KL, Jones EG, Kane SA (1983) Expression of GFAP immunoreactivity during development of long fiber tracts in the rat brain. Dev Biol 9: 317–336

Varon SS, Somjen GG (1979) Neuron-glia interactions. Neurosci Res Program Bull 17: 1–239

Vracho R (1982) The role of basal lamina in maintenance of orderly tissue structure. In: Kuehn K, Schoene H, Timpl R (eds) New trends in basement membrane structure. Raven, New York, pp 1–7

Waddington CH, Perry MM (1966) A note on the mechanisms of cell deformation in the neural folds of amphibia. Exp Cell Res 41: 691–693

Warren F (1917) The development of the paraphysis and pineal region in mammalia. J Comp Neurol 28: 75–135

Watterson RL (1965) Structure and mitotic behaviour of the early neural tube. In: DeHaan RL, Ursprung H (eds) Organogenesis. Holt, Rhinehart & Winston, New York, pp 129–159

Wessells NK, Spooner BS, Ash JF, Bradley MO, Luduena MA, Taylor EL, Wrenn JT, Yamada KM (1971) Microfilaments in cellular and developmental processes. Science 171: 135–143

Witschi E (1962) Development: rat. In: Altman PL, Dittmer DS (eds) Growth, including reproduction and morphological development. Biological Handbooks. F.A.S.E.B. Washington, DC p 306

Zeligs JD, Wollman SH (1979) Mitosis in thyroid follicular epithelial cells *in vivo*. J Ultrastruct Res 66: 288–303

Advances in Anatomy, Embryology and Cell Biology

Editors: F. Beck, W. Hild, W. Kriz, J. E. Pauly, Y. Sano, T. H. Schiebler

Volume 122

K. Y. Reznikov, University of Moscow

Cell Proliferation and Cytogenesis in the Mouse Hippocampus

1991. Approx. 80 pp. 30 figs. 7 tabs.
Softcover DM 98,– ISBN 3-540-53689-2

This research monograph reviews the results of the study of cell proliferation, cell death, neurogenesis and gliogenesis in the mouse hippocampus. The book presents original maps of distribution of mitoses and pyknoses in the developing Ammon's horn and dentate gyrus. It also gives an analysis of the location, age dynamics and origin of proliferating and dying cells. Data is given on how cell composition is formed. In the concluding section, the specific features of neurogenesis in the hippocampus and their possible relation to learning and memory processes are discussed.

Volume 121

P. H. M. F. van Domburg, H. J. ten Donkelaar, University of Nijmegen

The Human Substantia Nigra and Ventral Tegmental Area

A Neuroanatomical Study with Notes on Aging and Aging Diseases

1991. X, 132 pp. 38 figs. 4 tabs.
Softcover DM 110,– ISBN 3-540-52823-7

This book provides a comprehensive survey of the structure and fiber connections of the human midbrain, specifically of the substantia nigra and ventral tegmental area. The cellular and chemical architecture of these structures is analyzed and their fiber connections are discussed.
The role they play in degenerative diseases of the nervous system, such as Alzheimer's and Parkinson's diseases, is evaluated. Some functional and pathophysiological considerations are included.

Volume 120

L. J. Wurzinger, Technical University of Munich

Histophysiology of the Circulating Platelet

1990. VII, 96 pp. 42 figs. 9 tabs.
Softcover DM 68,– ISBN 3-540-52258-1

This volume closes the gap between knowledge of the platelet based on in vitro studies and knowledge of platelet hemostatic and thrombogenic function in vivo. An exhaustive review of the relevant literature, including recent ultrastructural and cell-biological studies, forms the bridge between basic research concepts and their implications for platelet function in the flowing blood.

Volume 119

D. E. Oorschot, D. G. Jones, University of Otago, Dunedin, New Zealand

Axonal Regeneration in the Mammalian Central Nervous System

A Critique of Hypotheses

1990. VII, 121 pp. 38 figs. 16 tabs.
Softcover DM 89,– ISBN 3-540-51757-X

Research findings reviewed include: regeneration in developing mammals and in submammalian vertebrates, the use of transplants and/or pharmacological treatments, in vitro studies on neurotrophic and neurite-promoting factors and their potential relevance to CNS regeneration in vivo, and in vitro studies on the types of glial cells that may be responsible for enhancing or suppressing axonal regrowth.